D1395186

Collaborative Relationships in Construction

This book is dedicated to

David, Pat, Simon and Amanda

Betty and Stanley

Collaborative Relationships in Construction

developing frameworks and networks

Edited by

Hedley Smyth
Senior lecturer in Construction and Project Management
Bartlett School of Graduate Studies
UCL

Stephen Pryke
Senior lecturer in Construction and Project Management
Bartlett School of Graduate Studies
UCL

A John Wiley & Sons, Ltd., Publication

This edition first published 2008

Blackwell Publishing was acquired by John Wiley & Sons in February 2007. Blackwell's publishing programme has been merged with Wiley's global Scientific, Technical, and Medical business to form Wiley-Blackwell.

Registered office
John Wiley & Sons Ltd, The Atrium, Southern Gate, Chichester, West Sussex, PO19 8SQ, United Kingdom

Editorial office
9600 Garsington Road, Oxford, OX4 2DQ, United Kingdom
350 Main Street, Malden, MA 02148-5020, USA

For details of our global editorial offices, for customer services and for information about how to apply for permission to reuse the copyright material in this book please see our website at www.wiley.com/wiley-blackwell.

Library of Congress Cataloging-in-Publication Data

Collaborative relationships in construction : developing frameworks and networks / edited by Hedley Smyth, Stephen Pryke.
p. cm.
Includes bibliographical references and index.
ISBN 978-1-4051-8041-2 (printed case hardback : alk. paper) 1. Construction industry–Great Britain. 2. Business logistics–Great Britain. I. Smyth, Hedley. II. Pryke, Stephen.

HD9715.G72C632 2008
690.068–dc22

2008017523

A catalogue record for this book is available from the British Library.

Set in 10/12 Palatino by SNP Best-set Typesetter Ltd., Hong Kong
Printed in Singapore by Utopia Press Pte Ltd

1 2008

Contents

Preface

We completed our first collaboration, *The Management of Complex Projects: a Relationship Approach*, at the end of 2005 and it was published the next year. We had not particularly planned a sequel, but this is it, having stumbled into the venture as a result of conversations with Stephen Brown from the RICS Foundation and Madeleine Metcalfe, our commissioning editor at Blackwell, at the annual RICS COBRA Conference, held in September at UCL in 2006. We are pleased it came about.

The book adopts the relationship approach explored in the first book, yet does so casting the net wider. The book pushes beyond obsessions with projects and project management as discrete fields of studies. The UCL Construction and Project Management team within Bartlett School of Graduate Studies has been engaged in research into the management of projects, following Professor Peter Morris' seminal work published in 1994. The so-called 'front-end' begins with strategies to address corporate needs, for which a project becomes the means to the solution. Indeed, the management of projects includes the project in use, that is, beyond handover and final account. We have taken a relationship approach, arguing that it is people that add value throughout the construction process; the manner in which people work together heavily influences and can determine the effectiveness and efficiency of the task-orientated tools and techniques that have been the focus of many articles and texts already.

The book probes the corporate entities of the client and contractor organisations, analysing relationships in frameworks and how these fit into a broader concern for programme and project management for clients and for contractors where value is being added. The book also pushes beyond the organisational boundaries into networks of relationships that involve a broad range of actors in the identification of resources that are levered to help meet expectations and secure financial reward.

The exploration provided comes in the form of a critique of traditional practices both conceptually and through evidence that highlights the constraints of many conceptions and practices. It draws attention, both directly and indirectly, to unconscious shifts amongst key project actors and analyses some emergent trends amongst players in explorations and transition towards new ways of working. Therefore, the book is providing a challenge to researchers and practitioners that will test preparedness to reform and modernise beyond current considerations of good practices. We are not posing the book as prediction – it's a matter of will and choice, but where there is a will we show there are ways. Nor is it normative for we do not prescribe particular panaceas, but provide a context for others to set out their own stall. This is all working towards improved understanding of the relationships to more effectively manage frameworks and networks through which projects are delivered.

We would like to thank those mentioned above and all those in our own networks who in different ways have influenced the way we have come to understand the management of projects.

H.S.
S.D.P.
London, UK

About the authors

Summary details are provided about each author, arranged in alphabetical order.

Professor Dilanthi Amaratunga is a Chair in Quantity Surveying at the School of the Built Environment, University of Salford.

Aaron Anvuur is a Lecturer in Construction Economics and Management at the Department of Building Technology, Kwame Nkrumah University of Science and Technology, on leave to The University of Hong Kong as a PhD candidate.

David Baldry is Associate Head of Teaching at the School of the Built Environment, University of Salford.

Jessica Chen is a Research Associate in the Centre for Interdisciplinary Built Environment Research, Faculty of Engineering and Built Environment, University of Newcastle, Australia, and teaches undergraduates in architecture and construction management.

Professor Charles Egbu is Professor in the School of the Built Environment at the University of Salford, and researches and lectures in Construction and Project Management.

Dr Richard Haigh is a Lecturer in Construction Management at the School of the Built Environment, University of Salford.

Kaushal Keraminiyage is a Research Assistant at the Research Institute for the Built and Human Environment, University of Salford, where he is currently conducting his PhD.

Professor Mohan Kumaraswamy teaches and researches in Construction Project Management at the Department of Civil Engineering, and is the Executive Director of the Centre for Infrastructure and Construction Industry Development, at The University of Hong Kong.

Dr Cynthia Lee is a Researcher in the School of the Built and Natural Environment, Glasgow Caledonian University.

Dr Kerry London is an Associate Professor in Architecture and Construction Management, in the Faculty of Engineering and Built

Environment, University of Newcastle, Australia, and is the Director of the Centre for Interdisciplinary Built Environment Research.

Gangadhar Mahesh researches on contract negotiations and procurement systems and is a PhD candidate in Construction Engineering and Management at the Department of Civil Engineering, of The University of Hong Kong.

Jim Mason teaches and researches in the field of Construction Law, Faculty of the Built Environment, University of the West of England. He is a non-practising solicitor and programme leader for Quantity Surveying and Commercial Management.

Dr Chaminda Pathirage is a Lecturer in Quantity Surveying at the School of the Built Environment, University of Salford.

Keith Potts is a Senior Lecturer in Quantity Surveying and Project Management at the University of Wolverhampton, and is Award Leader for the RICS accredited MSc in Construction Project Management.

Dr Stephen Pryke is a Senior Lecturer in Construction and Project Management, Bartlett School of Graduate Studies, UCL, and is Director of the Masters course Project and Enterprise Management.

Menaha Shanmugam is undertaking research at the Research Institute for the Built and Human Environment, University of Salford, where she is currently conducting her PhD.

Dr Hedley Smyth is a Senior Lecturer in Construction and Project Management, Bartlett School of Graduate Studies, UCL, and is Director of the Graduate Research Programme.

Leentje Volker is a part-time Researcher at the Faculty of Architecture, Delft University of Technology, in the Netherlands where she is currently conducting her PhD on Design Management. She is also a part-time Project Manager and Researcher at the Center for People and Buildings, an independent organisation founded by Delft University of Technology.

Foreword

This book represents an important step forward in the development of COBRA, the construction and building research conference of the Royal Institution of Chartered Surveyors. For the first time, we have taken the work presented at COBRA, which represents the very latest work of researchers from around the world, and encouraged researchers to think further about their work and present it within a unifying context, in this case the management of collaborative relationships and the management of projects.

This book is welcome and important for a number of reasons. Firstly, it has provided an opportunity to take ideas and concepts put forward by a range of researchers at the 2006 COBRA conference and to take them to the next level. It will enable these ideas to be discussed and debated more widely and, as a result, for the body of knowledge to be moved forward. It also represents the best in academe, with researchers from around the world, and from different stages in their academic lives, being prepared to work together to address issues of real-life concern to industry and society. Given the focus of the book, on collaborative relationships, this book is a credit to everyone who showed themselves prepared to be involved in this endeavour. Finally, it is in itself a valuable and important collection. In an industry which is often, and often unfairly, criticised for problems with the management of major projects, the insights contained in this book are timely and welcome.

Stephen Pryke and Hedley Smyth are to be congratulated on bringing together this book, which represents the first in what I hope is an annual series of books to emerge from the COBRA conference. They have set a standard that others must now seek to meet. I look forward to the next books in this series.

Stephen Brown
Head of Research
Royal Institution of Chartered Surveyors, London

Introduction

Managing collaborative relationships and the management of projects

Hedley Smyth and Stephen Pryke

Value is added to projects through people. Individuals and people working together as teams use the tools and techniques for managing projects. The development of front-end strategies and tactics, and the project execution phases, are only as good as the people behind them. For complex projects with high levels of uncertainty effective management of complexity and risk depends upon how well people work together. Therefore the management of relationships is important, yet much of the project management literature has focused upon managing the tools and techniques. In the book *The Management of Complex Projects*, Pryke and Smyth (2006) began to explore and redress this imbalance, developing a *relationship approach* to managing projects. This book on collaborative relationships takes the approach forward, considering relationships in the broader context, which in practice raises many issues that those managing projects are forced to address.

The *relationship approach* is complementary to other approaches or 'paradigms'. It is not a replacement. The four major paradigms are set out in Table I.1. This conceptualisation is inclusive of the 'management of projects' (Morris, 1994) and the project execution emphasis of 'project management' (PMI, 2004). It is also inclusive of 'critical management theory' employed in the research of projects, but recognises that critical theory currently tends towards seeking out particular phenomena as a challenge to the application of project management tools and techniques in contrast to most other conceptualisations, which try to impose rational order on a project environment of complexity and uncertainty. There is a legitimate place for both. A relationship approach acknowledges the context and conditions that can disrupt rational applications and produce particular, often unpredictable, outcomes or events. Yet, the approach also recognises that there are patterns of phenomena and activities that produce general, often regular, events and outcomes (Smyth and Morris, 2007; Smyth *et al.*, 2007). Indeed, one of the key themes of the

Table I.1 Paradigmatic approaches to managing projects (Pryke and Smyth, 2006; Smyth and Edkins, 2007; Smyth and Morris, 2007; Smyth *et al.*, 2007)

Traditional project management approach
An execution based approach that applies tools and techniques to impose rational order and efficiency. The approach has a strong emphasis on control and is task orientated. Scheduling tools and earned value analysis provide two classic examples, lean production and supply chain management provide two more recent examples. Other recent developments such as critical chain and performance management have brought an added social science and behavioural element, moving away from predominantly linear thinking and a simple cause–effect model

Information processing approach
An execution emphasis coupled with a wider consideration of stakeholders within the project coalition. The approach applies a technocratic input–output model of managing projects. It contains social theory and tends towards an efficiency focus, drawing upon economics and managerialist sociology to address information as a means of reducing uncertainty and improving attendant risk management. This paradigm tries to graft a more integrated approach onto linear task-orientated thinking, and human dimensions tend to be subsumed under the technocratic and managerial considerations. It is primarily based upon a simple cause–effect model, but takes account of emergent phenomena, especially additional information, to moderate the project strategy and implementation

Functional management approach
A strategic and 'front-end' emphasis, which is coupled with project execution to provide greater integration in what has come to be known as the management of projects. Organisational design and social theory provide a focus upon effectiveness as well as efficiency, embracing structures, open systems and processes, including supply chain management and other task-driven agendas, in pursuit of functional outcomes. There is a broader appreciation of human and organisational behaviour, indeed a range of people issues that goes beyond the task-orientated focus. Taking account of internal and external factors and the front-end provides broader definition of the project linking into business, strategy, portfolio and programme management, as well as issues such as learning, competency development and stakeholder analysis. The thinking is less linear, although research methods have not always recognised this

Relationship approach
Relationships add value, the tools and techniques being as good as the hands they are in. Relationships are seen as means to improve project performance and client satisfaction, achieved through relationships between people, between people and firms, and between firms as project actors that can be actively managed socially. It is based in social theory and tends to focus upon effectiveness. This approach draws upon a diverse set of research, ranging from inducing less adversarial behaviour through top-down, procurement-led measures that change market structures and governance to proactive behavioural management conceptualised and practised through relationship management; addressing the object–actor dichotomy as relationships are not only actor-to-actor, but actor-to-object too; and also incorporating perspectives from critical management and critical social theory. This approach is not a substitute for, but is complementary to others. As the approach is theoretically diverse there are a range of methodologies and methods used in practice. Typically thinking is non-linear, researchers seeking the general and particular outcomes, recognising the prevailing context and conditions

relationship approach is that the management of relationships is likely to achieve a better balance in managing both general and particular project factors, the known and the uncertainty, with less risk and greater success (Pryke and Smyth, 2006).

The relationship approach is also inclusive of 'relational contracting', which tries to change market structures and governance, typically through procurement initiatives in construction, for example partnering and supply chain management. Relational contracting is frequently applied on a project-by-project basis because many clients and most contractors have yet to embed and manage initiatives at the corporate level in order to consistently implement them across programmes and beyond the upper tiers of supply chains. Relational contracting seeks market and governance changes in order to induce changes in behaviour and management. Research shows that this 'solution' tends to leave responsibility for behavioural change at the level of individuals rather than through bottom-up, proactive management of behaviour (Smyth and Edkins, 2007). The relationship approach provides conceptual and practical scope for a proactive approach as well, for the development of effective relationships (for example using organisational learning, emotional intelligence, relationship marketing and management, and through trust – see Pryke and Smyth, 2006).

Definitions

Relationships are primarily between individuals from a socio-psychological perspective, and from a network theorist's viewpoint, between firms and indeed inanimate objects (computer terminals and railway stations for example)(Pryke, 2005). Individuals can have indirect relationships with organisations (see, for example, Chapter 5 by Anvuur and Kumaraswamy and Chapter 6 by Smyth), which are derived indirectly through the aggregation of experience and through reputation of the product or service and of the enterprise or organisation as a whole. Business-to-business relationships are therefore indirect in a behavioural sense, being conducted by individuals and teams, yet are direct through contractual agreement and financial exchange. Positive relationships tend to be *collaborative* in a commercial sense (which does not preclude friendship, although this is not a necessary requirement for collaboration). There are degrees of positive collaboration, which are explored in Section II (see especially Chapter 5 by Anvuur and Kumaraswamy).

Interpersonal relationships are not confined to organisations or relationships between organisations, but are conducted in *networks* that span boundaries. These are social, yet provide the means to support and channel technical and economic matters that feed into projects to deliver both basic services and added value. There are also networks within organisations (Pryke, 2005).

Framework is a term used in two senses in the book. First is the traditional conceptual sense, where a series of concepts are linked to help structure ideas, understand processes and provide a basis to organising action on the ground. Frameworks are not defined as theory although they may form part of or feed into a theory or theoretical approach. They are particularly useful in understanding how to manage projects as the subject does not have its own body of theory, drawing upon a range of disciplines. Second, and the most relevant to the subtitle of this book, is *framework* in the sense of framework agreement – series of projects linked together to provide continuity of relationships extending beyond project contract periods. For example a framework approach is used by many large clients for a programme comprising a number of projects. Although framework agreements have been formalised and some might say have become bureaucratic in their operation, Latham's vision of contractual commitment to future workload has not been feasible in practice. Client organisations rarely feel confident to predict *and effectively guarantee* future workload, preferring a 'statement of intent' type of approach, thus avoiding the litigation that would flow from the inevitable failure to accurately predict workload some years in advance.

Investments and procurement are distinct, yet linked. Investments may be organised as a *portfolio* of separate investment decisions. Investments may be policy related, geographical, business unit/divisional and product related. For example an investment conglomerate may have corporate bonds, development property, hotel investments and premises used for manufacturing as part of its portfolio. Investments may be organised into sub-categories for implementation as *programmes*, such as investments that share project procurement. Parts of programmes may be bundled into *frameworks/framework agreements* with shared attributes whereby bundling increases capability to leverage lower costs or added value.

For some organisations their investment portfolio, programmes and frameworks may be synonymous, yet there are clearly classification differences. The construction client has been used as an example for such classifications, as few contractors organise their portfolio of projects as programmes strategically and tactically supported from the centre or head office in the way portfolio and programme management implies. Yet the demands from clients for improved value and services may increasingly challenge this approach. Contractors may need to frame their business into programmes (rather than procurement silos – see Smyth, 2006), providing investment and support from the corporate centre of the contracting enterprise.

Scoping frameworks and networks using a relationship approach

In scoping the relationship approach the social environment is depicted according to Figure I.1.

To date, primary emphasis has been placed upon 'The Project', external stakeholders, supply chains and clusters in delivering value added and added value through relationships to clients (Pryke and Smyth, 2006). Particular emphasis has been put upon the project and how individual organisations, such as the contractor, can manage their projects more effectively and efficiently (see the relationship model presented in Pryke and Smyth, 2006:34–41).

The focus continues here, yet the net is cast wider. The broader social network is included. This network can be viewed as internal and external. First is the external network (Figure I.1) where 'independent parties' contribute directly and indirectly to the way that projects are managed, yet are not subject to contracts for the supply of goods associated with the immediate project supply chains and clusters and are not part of the project design team – they are external to the project coalition, as defined by Winch (2002). Second are the internal networks that are part of the coalition and maybe part of the decision-making unit (DMU) nor other project actors within organisations involved (Smyth, 2000; Pryke and Smyth, 2006). One example picked up in chapters in this

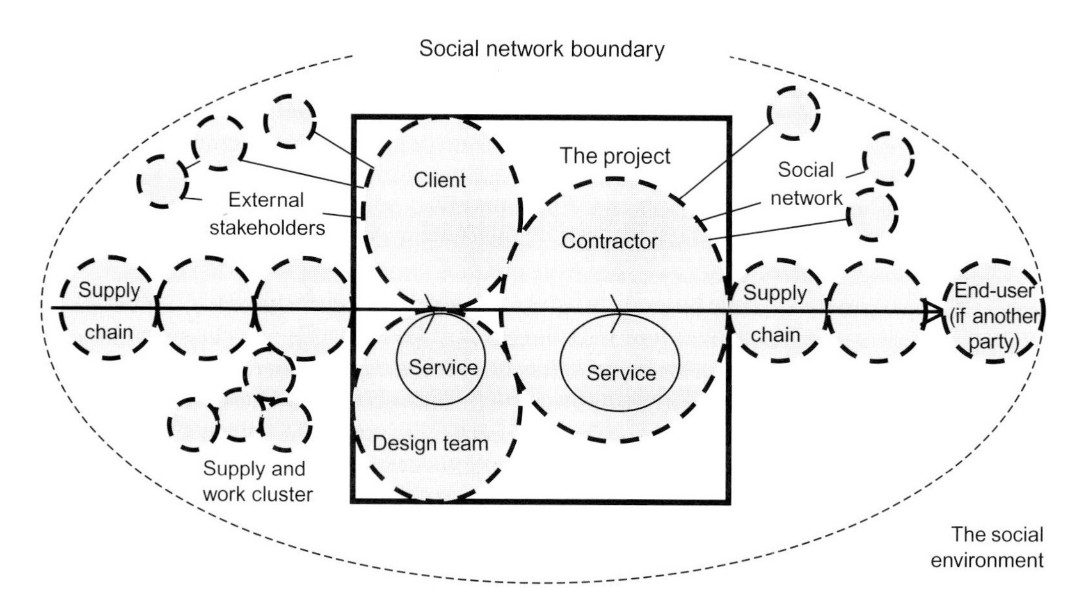

Figure I.1 The project environment: total social context and structure (Pryke and Smyth, 2006:31)

book is the role of public organisations, policy effects and implementation gaps within their internal networks that impinge upon effective project operations (see Chapter 7 by Haigh *et al.* and Chapter 9 by London and Chen).

The net is cast wider in a further sense. As indicated, the corporate–project interface is important for investment and managing consequential costs that are incurred during the life of projects managed as frameworks within programmes. The contractor and consultant motivation for adopting a relationship approach is to meet the increasing demands of the most sophisticated clients and complex projects on the one hand, and to increase repeat business for these clients as a minimum and preferably to earn premium profits where possible (conceptually on the added-value aspects in particular).

Increased certainty of repeat business arises where clients create framework agreements and appoint panels of consultants and contractors. The consultants and contractors receive repeat business through negotiated appointment or tendering against other panel members. The development of such frameworks forms part of the client management of project strategy for each project, which constitute part of the programme of projects. Thus, frameworks are part of client programme management. *Programme management* is an area of increasing interest and focus for projects (e.g. Gareis, 2004; Thiry, 2004; Partington *et al.*, 2005). Construction clients have been quicker than contractors to adopt programme management methods to aid the management of projects. Contractors have tended to manage projects on a project-by-project basis, even where continuous improvement agendas are being pursued and where framework agreements are in place for partnering and for using integrated teams (Olayinka and Smyth, 2007).

It does not require formal framework agreements for contractors or consultants to manage their portfolio of projects as programmes perhaps defined by procurement route or market segment (as opposed to divisional silos organised by procurement in order to minimise necessary support – see Smyth, 2006). Growing numbers of contractors that have been transitioning from relational contracting towards relationship management have been adopting a programme approach to client and project management (e.g. Smyth and Fitch, 2007).

The consequence is that a broader outlook is considered here: corporate–programme–project–client interfaces within a network of social relationships (see Figure I.2). It is this broader outlook which provides the focus for the book and the contextual locator for author contributions. Hence, collaboration is more than intention. It needs support within the contracting enterprise and beyond organisation boundaries. Relationships provide the vehicle for collaboration yet these need organising, strategic support and investment from the corporate centre to be organised formally or informally through programme management.

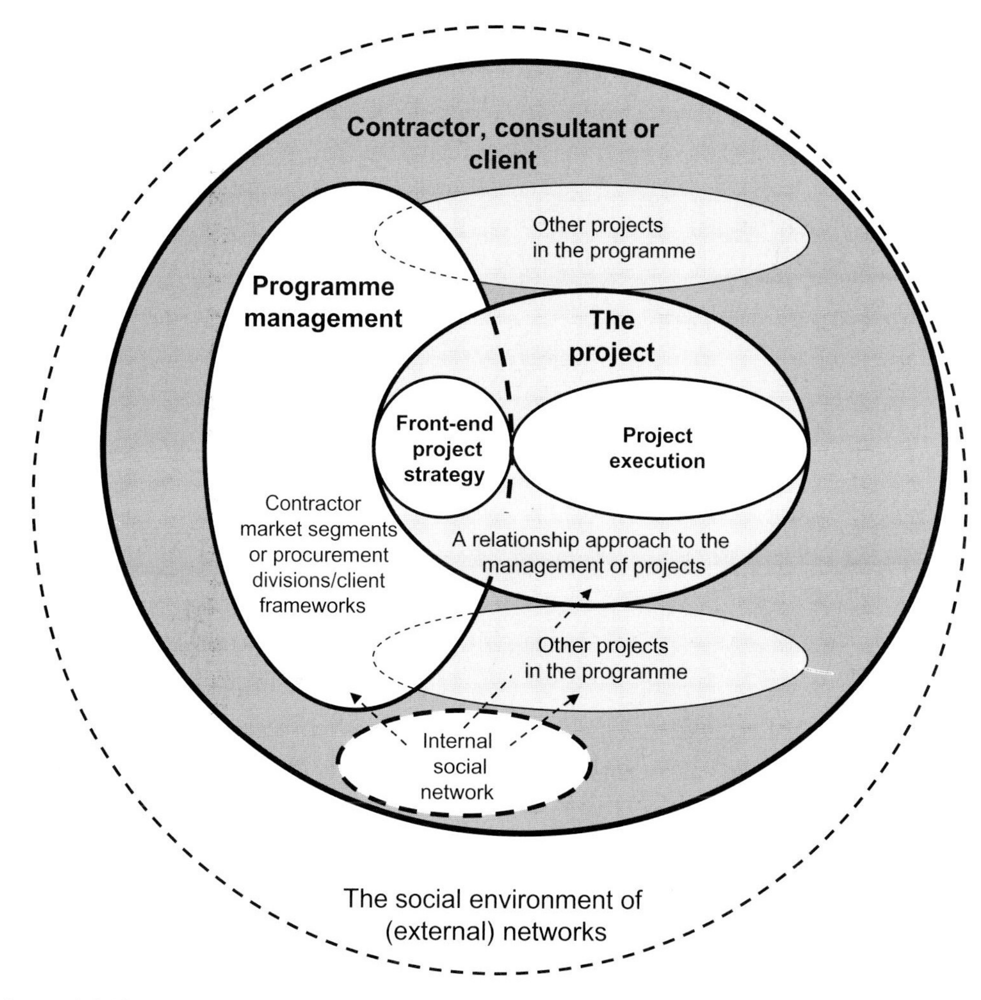

Figure I.2 The corporate–programme–project–client interfaces within a network of social relationships

Aim and objectives

This broad focus of collaborative relationships in a social environment of networks concerns current practice and concerns potential for future development. Whilst this potential can be read as prescriptive or normative – what ought to be applied – this is not strictly the way it is presented in this book. There is reporting of current application *and* there is a conceptual exposition of what is possible respectively. Rather than prescription, the conceptual exposition of what is possible is a challenge for the construction industry. Reform agendas around the world have yielded some results (e.g. Beach *et al.*, 2005; Olayinka and Smyth, 2007), but have also fallen short of expectations (e.g. Cox and Ireland, 2006; Green, 2006; Smyth and Edkins, 2007). A realistic assessment suggests a range of responses:

- Some enterprises have not responded for a range of reasons.
- Some enterprises have responded in rhetoric, fashioning their language to the agendas without any fundamental change in behaviour.
- Some enterprises have responded on a project-by-project basis in response to client demands – typically in forms of relational contracting.
- Some enterprises have decided to pursue such agendas on a more consistent basis, if not in total within procurement divisions.
- Some organisations are transitioning from relational contracting to forms of relationship management (cf. Bresnan and Marshall, 2000; Green, 2006; Smyth and Edkins, 2007; Smyth and Fitch, 2007; Olayinka and Smyth, 2007).

Some of these responses are picked up by authors here (e.g. Chapter 1 by Mason, Chapter 2 by Potts and Chapter 4 by Kumaraswamy *et al.*), particularly opportunities to transition to more proactive ways of developing and managing relationships – cooperation and collaboration (e.g. Chapter 5 by Anvuur and Kumaraswamy and Chapter 8 by Volker). The issue with transitioning from relational contracting (changes in market structure and governance) to relationship management (direct management of organisational behaviour) is simply that the law of diminishing returns kicks in for indirectly inducing behavioural change, and hence constrains scope for improvement, under relational contracting. Further opportunities for collaboration are opened up by developing relationship management within the supply chain (Pryke, 2008 forthcoming) and in the enterprise, for example core competencies in construction (Smyth, 2011 forthcoming). Relationship management with enterprises and across the network offer scope for continuous improvement. There is no conceptual imperative for construction enterprises to adopt such approaches, but the conceptual challenge is a test to both the client agendas for industry and the progressiveness of construction enterprises. If this is all fashion that will fade until a new one is identified (Green, 2006), then the sector will ultimately fail the test. If part of the industry responds and creates new opportunities to add service and product value then the challenge will have been met to the extent of the response in breadth and depth.

For proactive development in the management of collaborative relationships, construction enterprises need greater support from the corporate centre than has typically been the case. Investment is required and costs incurred to yield a net return, the relationship profit (see the relationship model in Pryke and Smyth, 2006:34–41). This changes the corporate–project interface. The main *aim* in this book is to focus upon the corporate–programme–project–client interfaces within a social network using a relationship approach. The *objectives* are to explore the following challenges:

- Continuous improvement to project performance whereby construction contractors provide support from the corporate centre of enterprises that goes beyond the traditional bounds of activity, for example estimating and procurement.
- Clients will continue to drive sectoral change, requirements becoming more demanding with expectations that added value will continue to increase over time.
- Core and repeat business clients have investment programmes which are organised into formal or informal frameworks for the allocation to panels of pre-selected contractors on a competitive or negotiated basis, contractors needing to maintain their position on such panels through providing continuity and consistency of service across series of projects in order to add service value.
- The most responsive contractors will increasingly need to transition from relational contracting to forms of relationship management.
- Contractors will increasingly manage their projects as programmes, perhaps identifying market segments of added relationship value to create several programmes that make up the portfolio of their total contracting activity.
- Improvement to client–contractor relationships and the drive to improve project performance is strengthened through the recognition that designer, supplier/sub-contractor relationships are vital in delivering added product and service value. This recognises that firm boundaries are perforated as knowledge as well as products are delivered through broader social networks that span beyond project boundaries, organisational boundaries and complete supply chains and clusters.
- Constructors and professional service providers need to exploit the opportunity the long-term collaborative relationships present to innovate and drive down clients out turn costs, as well as costs in use.

A relationship framework

Managing projects is traditionally seen as execution based (see Table I.1). This is encapsulated in bodies of knowledge, such as the Project Management Institute *PMBOK® Guide* (PMI, 2004), the Association for Project Management's BOK (APM, 2006), the popular PRINCE2 accreditation developed for public sector projects, or the RICS list of competencies associated with their project management faculty. Whether a formal body of knowledge is applied for managing projects, a formal set of tools and techniques derived externally from or internally within the contracting enterprise, the aim of management is to bring more order, guidance and direct control. The objective is successful delivery of the service, hence the successful delivery of the 'product' – the completed project. Yet all projects carry considerable uncertainty; hence risk (Winch, 2002). Uncertainty arises within the project boundary (social and physical, hence technical too), the definition of which will vary as to whether a project management (execution emphasis) or management of projects view (including a front-end strategy) is adopted. Uncertainty also arises from a broad range of stakeholders, including the project coalition (Winch, 2002) and other external factors and forces that extend into the social network and beyond into a broader social and political environment.

It is suggested that project managers spend a great deal of their time managing these uncertainties. These are precisely the issues that are not easily subscribed by the bodies of knowledge or the range of project management tools. The management of projects approach helps overcome some of these, and, therefore, frameworks and programme management add further support. Yet, such measures do not overcome all uncertainties. It is these uncertain conditions that give rise to particular outcomes that are typically out of line with the project and programme objectives if left to their own devices. This is a primary reason for management. This is why it is typically acknowledged that projects are context specific (e.g. Griseri, 2002; Morris and Pinto, 2004; Smyth and Morris, 2007). And yet, context specificity is frequently overlooked in the way that managing projects is socially constructed in the bodies of knowledge and formulated in practice for particular projects within and between the enterprises responsible for management. Indeed, research frequently replicates the same problems. Research methodologies and methods applied typically seek either to identify and explain the general or to identify and interpret the particular. They largely fail to create a balance between management generating generally prescribed outcomes and the particular context that affects whether and how these outcomes are achieved (Smyth and Morris, 2007). Management may be able to bring to bear the range of knowledge and tools to maintain the desired outcomes, but collaborative relationships within and beyond teams provide means to engineer successful outcomes. It may also be the case that different solutions are applied in these ways to produce particular outcomes which are successful, in contrast to those originally prescribed in the formal knowledge base and sets of tools. They may also fail, which is a different sort of (particular) outcome (Figure I.3).

Relationships therefore become a key focus, not only for effective application of the bodies of knowledge, the management of projects and

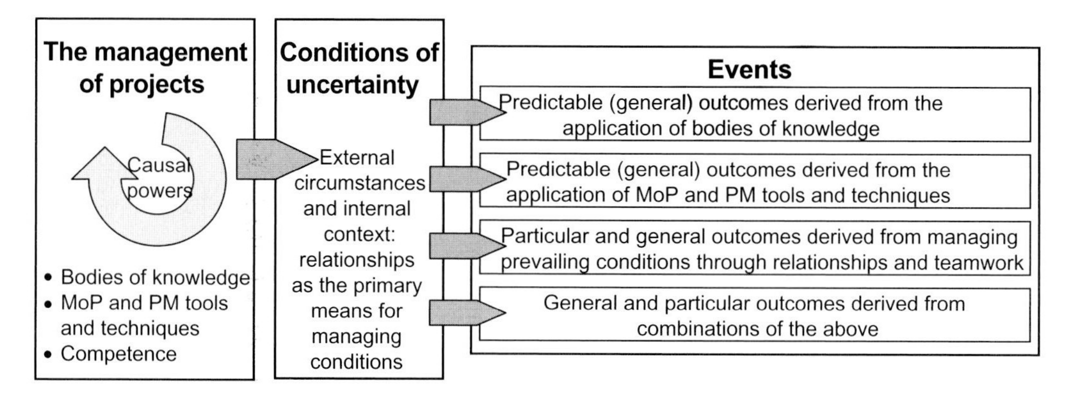

Figure I.3 A critical realist causality model for the management of projects within a relationship approach

project management tools and techniques, but also for managing the contextual conditions because knowledge and techniques provide insufficient clarity and guidance for controlling these factors. This way of viewing methodology and methods is derived from critical realism (e. g. Sayer, 1992; Danermark *et al.*, 2002) and is applied to project management (Smyth *et al.*, 2007).

This analysis raises the issue of the transitory nature of social relationships and networks. One of the main aims of management is to produce stability through managing uncertainties out of the environment and providing stability by controlling the factors within their powers. The presentation in Figure I.3 identifies the ways in which this operates in practice. Relationships are clearly a critical factor in delivering successful projects, yet paradoxically relationships are infrequently managed. Management is conducted on a task-orientated basis across many projects (Handy, 1997). This means that behaviour, whilst guided by organisational culture (and project culture where projects are complex and have a long duration with relatively stable teams in terms of key personnel), is left to the responsibility of individuals (Smyth and Edkins, 2007). The traditional adversarial behaviour has been indirectly addressed through relational contracting, yet moves to address it directly through relationship management, the development of emotional intelligence or another behavioural core competency have been tentative to date in construction. This paradoxical situation is beginning to show selective signs of change and this book helps to unpack the potential for such change – not necessarily in a prescriptive way but as a challenge to the construction industry.

Managing relationships is of potential importance for improving collaboration, yet it must also be acknowledged that all relationships are conducted contextually too (Gilligan, 1982), and thus form part of the conditions for a project, framework and network. Frameworks perpetuated through alliances and networks are contextual too. However, frameworks endeavour to remove some of the contextual uncertainties and complexities, drawing the project into the management of projects domain. Such frameworks are typically organised as formal contracts and alliances as part of investment programmes of clients, or contractors organising projects into programmes, supported by the corporate centre and a wider social and technical network.

Networks can be formal or informal and may be transitory or unstable, hence a challenge to manage. Whilst influence and power play important parts in structuring networks they are founded upon self-selection where some mutuality of benefits flowing must be forthcoming. Adopting a network approach towards the mapping and understanding of relationships is becoming increasingly important (Pryke, 2004) and collaborative relationships provide a particularly rich subject area for this type of approach. Network analysts see the network as the population, rather than being hampered by obsessions with organisational boundaries and hierarchies or vague allusions to 'atmosphere'. The environment, in which we live, work and complete projects, is populated with nodes between which information, money, influence and a wide range of other types of relationships exist and can be created (and terminated). Neural networks within the human brain similarly comprise a large number of nodes which connect momentarily, sometimes frequently, sometimes possibly uniquely, in order to achieve a particular task. If all of the nodes within the brain were permanently connected nothing would be possible because of excessive flows between the nodes. Within a project context, understanding the way in which nodes are connected – the nature of linkages between nodes and the role that each node performs – is important. Being able to classify the characteristics of the network nodes and the nature of the linkages between them opens up the possibility of understanding project networks and reproducing and managing the most effective formations in future project coalitions.

If it is accepted that there is value in thinking outside of the organisations within which project functions are delivered, then we open up the possibility of acquiring a much deeper understanding of what happens in project teams. Traditional methods of project management are associated with a number of very specific techniques for managing projects in a structured way and these are mentioned above. For complex projects and those where there is an element of innovation, we need to be thinking about the project as a system (Walker, 2007); indeed to be effective in dealing with the complexities of all but the smallest and most simple of construction or engineering projects, we need to be conceptualising the project as an 'open system' (Ackoff, 1981), responding to the needs of the clients and a large range of stakeholders continuously from conception to completion; perhaps never achieving completion in the case of the most complex of projects. Yet the practitioner or academic in search of understanding and analysis of these open systems is traditionally short-changed when it comes to analytical techniques. Networks provide insight here, through the application of social network analysis (Pryke, 2005; Pryke and Smyth, 2006; Ruddock and Knight, in press).

In this sense, frameworks are used to recognise and draw together key network actors and linkages. Programmes are trying to link projects into a wider set of organisational issues of strategy and decision making to aid their management and hence portfolio management. As such, this recognition is operating around the most central nodes and strongest links. Having considered the density and how practitioners apply frameworks to aid management, how the frameworks are actually managed can be articulated by analysing the network configuration along with the comparative centrality of the project actors. For example, this can be used in relation to supply chain leverage (Cox *et al.*, 2006). Identifying and managing flows through the network concerns social capital, knowledge and technical resources that are related to path length – the distance in terms of other actors through which resource flows occur. Some actors provide what might be referred to as a controlling or gatekeeper function; yet others act as catalysts or perhaps brokers and these actors can be fundamentally important in achieving successful projects. Networks aid the understanding of how people acquire knowledge within projects and transfer it to other projects; how they know what they know and how they acquire custom and practice through what Wenger (1998) referred to as *Communities of Practice*.

Understanding these networks will commence the process of operationalising much of the conceptual material that is discussed later in this book. For example, supply chains are seldom linear, typically being clusters organised as complex networks of human relationships functioning within a framework of legislative, contractual and incentive-based linkages. The multi-faceted networks constitute the supply chain carrying out its functions in relation to the project, to the programme and to much broader strategic roles.

Within the structure and operation of networks, individual relationships can be managed using relationship management principles. Some of the dimensions of relationship management have been explored elsewhere (Pryke and Smyth, 2006), and these are explored further, for example breaking new ground in relation to trust (Chapter 6 by Smyth). How far proactive management of behaviour is taken will depend on many factors. It would be possible to develop behavioural codes of conduct that form part of the systems and procedures specified by organisations for employees to conform to in performing their function and undertaking specific tasks. This might be considered too controlling or dictatorial in some organisations and the benefits derived would be offset by employees 'voting with their feet', yet enterprises are diverse in leadership and management styles and tend to attract employees who feel comfortable with particular styles (see for example Handy, 1997; Robbins, 2003; Johnson *et al.*, 2005).

One way of considering this approach to collaboration, specifically through networks and frameworks, is to observe that past research and rationalisation of practice has artificially tried to shoehorn activities into constrained management silos. This chapter contributes towards breaking out of such silos and encouraging wider perspectives. The subsequent chapters in this book show the broader dimensions frequently in operation in networks that penetrate projects *per se*, cross boundaries and intersperse both private and public sectors. Therefore, the book explores projects through a relationship approach. In so doing it is arguing that adding project value will increasingly be determined by proactively managing relationships:

- Within organisations, including across programmes.
- Across organisations, including frameworks.
- Across networks, using investment in and processes for relationship management as a means to draw actors together and deliver increased value.

Overview of chapters

Section I concerns collaborative relationships in contractual frameworks, exploring the practical scope and conceptual limits of collaboration within frameworks and programmes. Initiatives for continuous improvement based upon relational contracting are scoped and the potential for further and future development is addressed. In **Chapter 1** specialist contractors and partnering are examined by Mason. The research reports that main contractors rarely manage beyond the first tier, although some sub-contractors take it upon themselves to manage their suppliers on an informal basis through symbolic action, norms and financial controls. The value of this chapter is the clarity with which relational contracting issues are explored in supply chains, showing some of the potential, but also many restrictions, of current practices. The scope for further improvements in performance is presented, however, the evidence is clear that not all types of firms wish to be responsive and proactive. The implication is that continuous improvement and performance improvement via collaborative relationships will be stratified within the market and hence an important strategy issue for the future. One consequence may be a refining of networks with greater specialisation of actor and types of relationship in developing supply chain management.

Change in the quantity surveying (QS) profession, which is explored in **Chapter 2**, is part of sweeping changes that the construction industry has been embroiled in, which continue today. Some changes are initiated within the sector and some are stimulated or imposed from outside. Potts examines whether the QS profession is master of its own destiny or is undertaking change in response to client demands through the case study of Terminal 5, London Heathrow Airport. The main thrust is the change from cost monitoring towards cost management within a lean approach. The contribution of this chapter is analysis of the subtle shift of the QS from predominantly that of financial monitoring and reporting to a management role of adviser, facilitator and financial manager.

Lee and Egbu examine client project requirements and project team knowledge for refurbishment projects in **Chapter 3**. Client satisfaction, client project requirements and knowledge are the key themes. Knowledge, mainly tacit knowledge, accumulates the longer people have worked on a project and in the refurbishment sector. The focus is upon delivering increased client satisfaction from a greater identification and application of knowledge within project teams, gleaned through project experience and from a broader network too. In combining the management of knowledge with the management of projects this chapter takes an informal approach to developing greater collaboration supported by initiatives from the project team and their internal social networks.

In **Chapter 4** Kumaraswamy, Anvuur and Mahesh explore contractual frameworks and cooperative relationships. The paucity of conceptual thinking regarding contractual frameworks from an economic and management perspective is analysed. Transaction cost analysis provides a conceptual starting point. The chapter distinguishes between the transaction analysis provided by Williamson (e.g. 1979) and the more social analysis of Macneil (e.g. 1974). This misunderstanding has led to a misrepresentation of transaction costs. It opens up the analysis to approaches based around organisational behaviour, especially the conceptual approach of relationship management provided by others from the relationship marketing perspective (e.g. Gummesson, 2002). It also opens up the concept of frameworks in a broader sense for application in construction. The value of this chapter is that it conceptually shows the limits of current practice and begins to scope other approaches to improving performance. This develops the management of projects approach (Morris, 1994), linking it to the relationship paradigm for managing projects. It highlights the need for developing the corporate or organisational levels of management rather than from the project level *per se* as a means for developing and supporting service delivery.

Section II concerns collaborative relationships and conceptual frameworks within internal and external networks of relationships. The section particularly builds upon **Chapter 4** at the end of **Section I**. The conceptual scope of collaboration within frameworks, and projects constituted into programmes, develops the theme. The behavioural concepts for performance improvement are considered. In **Chapter 5** Aaron Anvuur and Mohan Kumaraswamy provide a conceptual appraisal of collaboration through cooperation and how improvements can be generated for theory and in practice. To date the basis for predicting linkage between cooperation and project performance has remained typically logical and conceptual rather than empirical. Cooperation has lacked conceptual and definitional clarity, confounding different levels of analysis. Socio-psychological factors that determine an individual's cooperative behaviour in construction need articulation, which the authors explain and develop. How socio-psychological factors can be affected by economic incentives is considered, and the chapter makes an important contribution towards this issue. Construction research needs to move on from *ad hoc* classification systems and taxonomies to comprehensive conceptual frameworks and theory-based systems. An in-depth analytical distinction is made between collaboration and cooperation. The chapter outlines the theoretical domain of cooperation at the level of the individual managers and identifies the theory-based dimensions of the cooperation construct as the bases for a metric for specific goal alignment, collaboration being more general and contractual.

The chapter therefore contributes a strong conceptual basis for collaborative relationships. The conceptual basis is located within a front-end management for projects approach. It is relationship based. There are also methodological implications to be drawn from the chapter, based upon contextual factors that inform behaviour, and from behaviour itself being contextual. The chapter also shows that improving collaboration through cooperative measures requires application to be undertaken with investment and incentives determined at the corporate level, not simply at the level of the project, certainly at a programme level, yet also for the whole project enterprise or organisation.

Chapter 6 takes a further analytical view of developing trust. Smyth reviews previous frameworks for conceiving trust. An amended framework is provided that takes greater account of internal and external factors of influence and power in relationships. Aspects of moral philosophy are injected into the analysis and applied to the concept of the moral economy. Therefore, the chapter develops the concept of trust in the context of collaboration within organisations, across organisations and in a network anchored in the moral economy. Context is therefore addressed and it is argued that trust analysis is methodologically located within critical realism for effective analysis. From this analysis the chapter moves towards identifying some behavioural principles that might underpin a behavioural code of conduct for developing and managing trust.

There is a tendency in construction to view a fashionable field for a period and then move on, so the value of this chapter is to retain a focus on trust and show that there remains conceptual and applied scope for identifying and managing trusting behaviour. This chapter is firmly located within the relationship approach and develops a theme of morality, especially the ethics of care and the moral economy, and methodologically in terms of critical realism. The outcome is that management strategy requires investment in the culture, systems and procedures and organisational behaviour at the corporate level in order to effectively introduce trust as a behavioural competency into projects, programmes and project framework agreements.

Section III develops the content towards collaborative relationships and networks. It describes how frameworks and programmes dovetail and develop into networks. Networks as social relationships go beyond the boundary of firms and organisations. In **Chapter 7** infrastructure lifecycles and disaster mitigation are considered by Haigh, Amaratunga, Keraminiyage and Pathirage. The role of government, NGOs and the construction sector in contingency planning for disasters provides the focus. Collaboration is considered at a high level conceptually with profound practical implications, linking through to policy and administrative networks that are governed in different ways to those responsible for project delivery. One way of looking at this issue, therefore, is to assume that the level of a 'disaster' is not simply the product of natural outcomes, but is also a social outcome – the consequence of inadequate provision. The contribution of the chapter is to demonstrate that front-end policy and project strategy are needed for disaster planning in order to provide social resilience to natural events. It is a government, NGO and private sector issue which requires policy coordination, greater inter-organisational collaboration, which can be achieved through formal mechanisms, yet also requires informal networking and support to ensure effectiveness.

Volker addresses early design management in architecture in **Chapter 8**. Selecting partners for and value judgment in design decision making and development provide the empirical focus. The chapter looks at how the project manager is appointed in the design phase with a strategic rather than execution orientation in this study. The emotional connection through the design is explored, and the manner in which this is used. As a result relationship commitments are forged, and the tools of selection for appointing a design practice/team mobilised. The analysis draws out the behavioural issues, 'the people side' of project and design management. Once more the role of the public sector is explored as a key stakeholder, albeit at a different level compared to **Chapter 7**, and the chapter shows how the selection process is both a rational and a subjective one. The subjective includes the application of intuitive decision making due to the complexity and interplay of issues and interests. Key players are taken into account as important stakeholders, not only the objective and subjective criteria from their own direct functional roles, but also through wider considerations within a network of stakeholder interests.

Chapter 9 also considers the role of government and how both policy and policy implementation gaps within and between administrative areas of responsibility play important parts in determining project outcomes. London and Chen focus upon infrastructure life-cycle supply chains, in their analysis, addressing difficult and complex issues that present many practical barriers to formal resolution, yet implicitly acknowledging the constraints of public accountability in reconciling problems and overcoming the barriers through more informal means.

The contribution of this chapter is derived from the execution stage of the project and the implications of front-end issues in supply chain management. This includes the absence of clear guidance or guidance that can be absorbed into practice. It also concerns the public–private relationship, not simply at the client–supplier interface, but also in the dynamics of the policy environment and the resultant diversity of behaviour arising from diverse perceptions and implementation gaps in the policies. This chapter therefore raises important substantive issues with strong network considerations. It also raises methodological issues.

A typical result of strategic decision making of transactional industries is the minimal expenditure on training and personnel development, culminating in an inward focus. **Chapter 10** takes an outward focus at the construction workforce, particularly the professional levels, and depicts the low representation of women in construction, comparing to and contrasting with the experience of women in the medical profession. Amaratunga, Shanmugam, Haigh, and Baldry argue that having a balanced representation of people in the workforce is beneficial for society and effective working. Gender issues are one aspect of having a balanced representation, and probably an important aspect of promoting greater collaborative working. In general the chapter considers the image of the sector from a gender perspective. It particularly shows how image and status in the medical profession has enabled change in the role and employment of women. The chapter shows the differences between the two sectors, but also demonstrates the lessons that can be adopted in construction to improve the role and employment of women in construction, especially the professions, as an exemplar.

The relationships that have dominated and still dominate construction in terms of a male culture and chauvinism, including blame, contrast with the facilitating and nurturing roles women tend to adopt. The chapter contributes insights as to how construction can benefit from such inputs in changing its culture as well as meeting current and anticipated shortfalls in skills. The chapter also serves to highlight the need to overcome the risks of being inward looking, that is, being isolated and failing to glean lessons from a wider network of practices that are pan-sectoral.

In the **Conclusion** a summary and recommendations for research and practice are provided, covering each section for improving future performance through conceptual and practical application of collaborative relationship working, especially within frameworks and networks. The need for proactive management of relationships rather than reliance on reactive market and governance procurement induced changes is set out.

Mapping the chapters on to the relationship framework

This chapter has set out a relationship approach, particularly concerning frameworks and networks, and how these feed into programme and corporate management of enterprises and projects. To an extent, the overview of the chapters shows how they individually contribute to this approach. A simple mapping exercise will add further clarification to this description and this is provided in Figures I.4, I.5 and I.6, which map individual chapters against Figures I.1, I.2 and I.3.

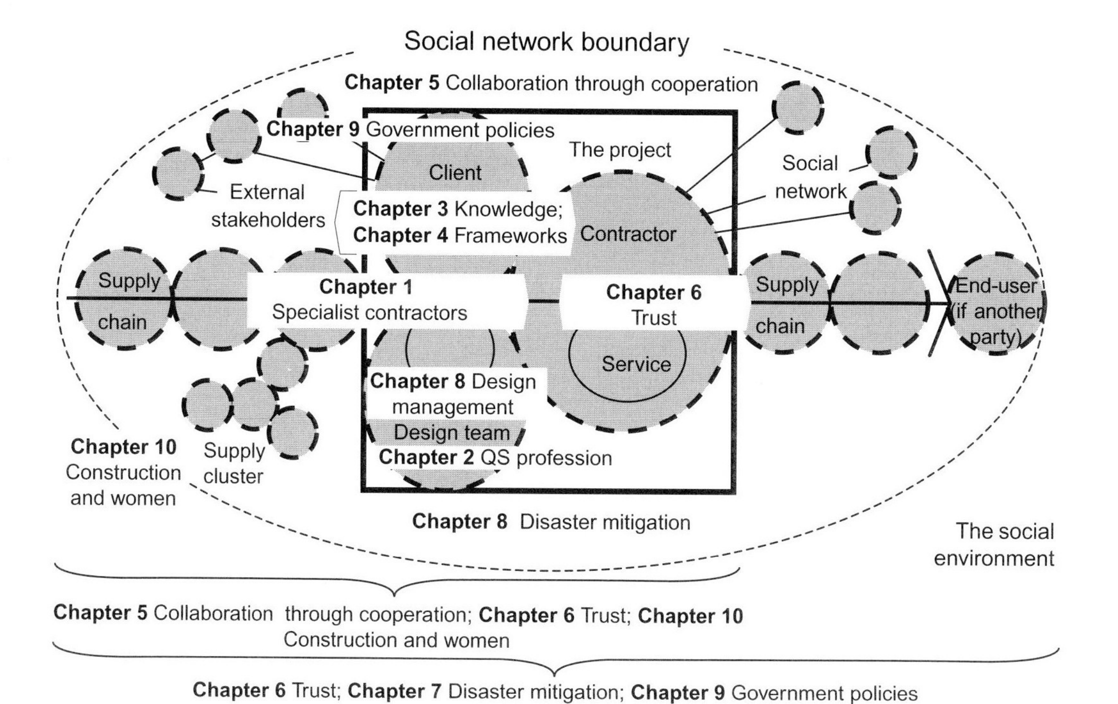

Figure I.4 Mapping the contributions against the total social project context and structure

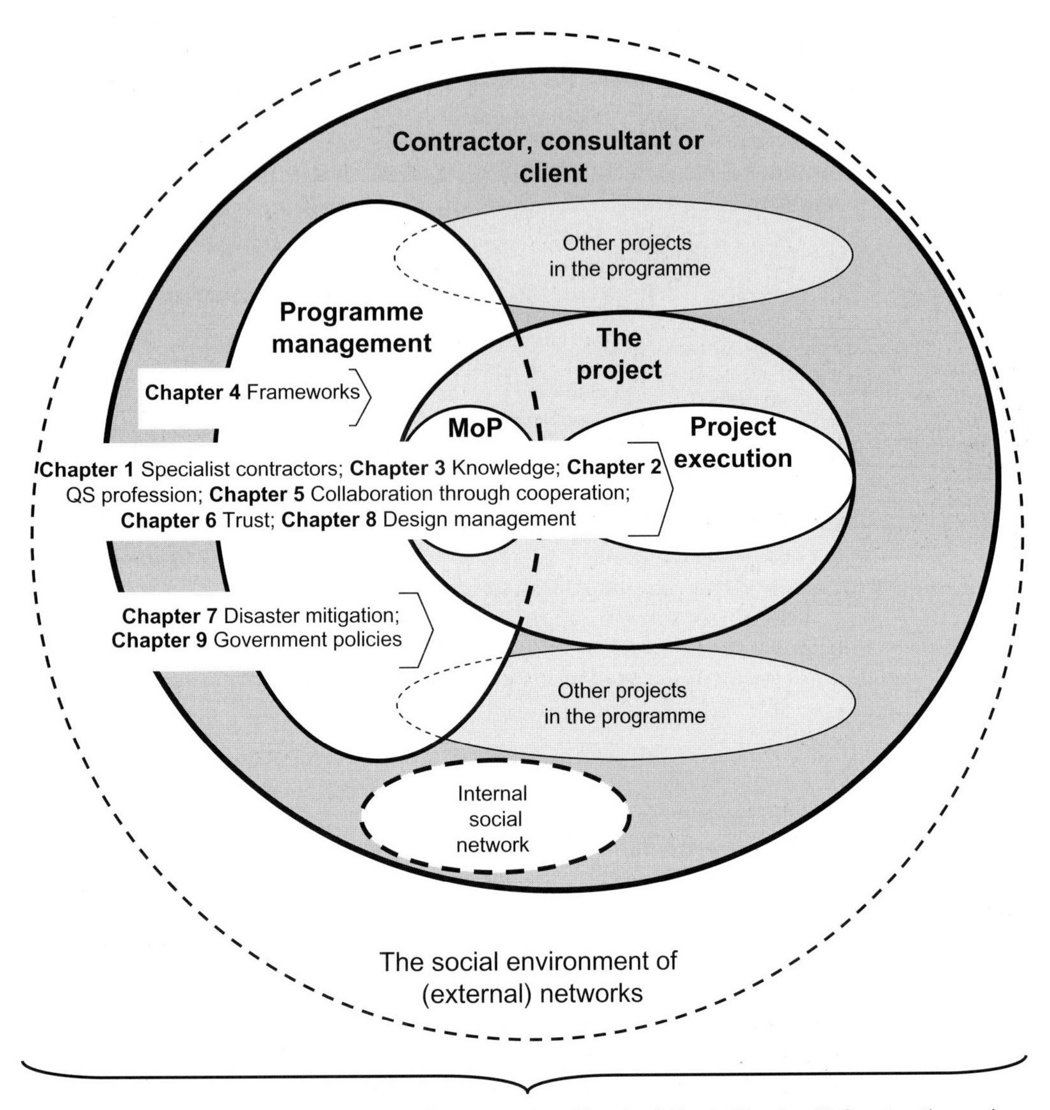

Figure I.5 Mapping the contributions against the corporate–programme–project–client interfaces within a network of social relationships

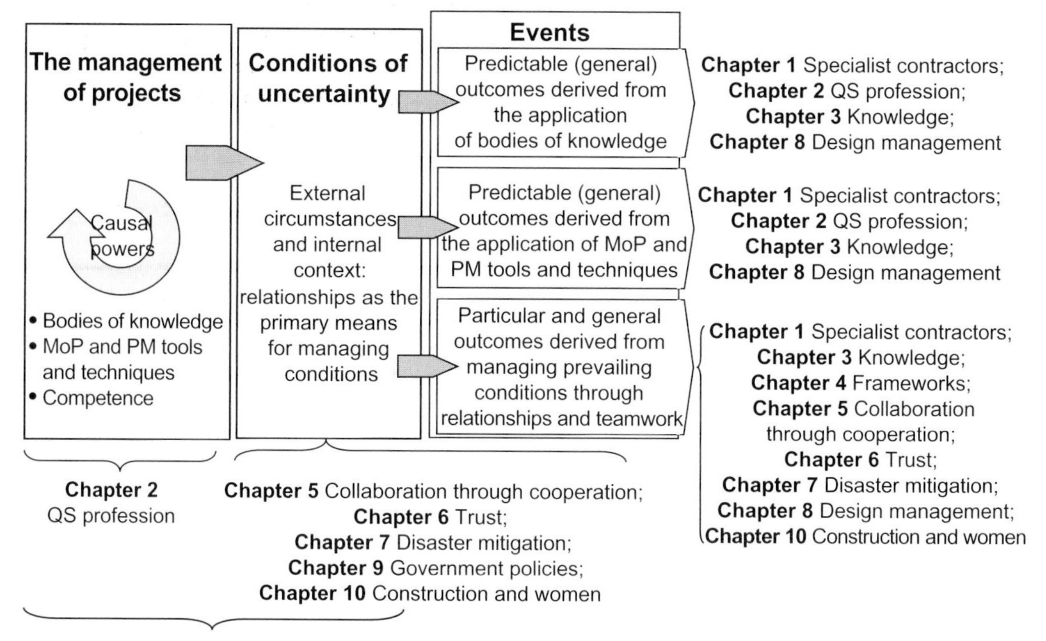

Figure I.6 A critical realist causality model for the management of projects within a relationship approach

References

Ackoff, R.L. (1981) *Creating the Corporate Future.* Wiley, New York.

APM (2006) *Project Management Body of Knowledge.* Association for Project Management, High Wycombe.

Beach, R., Webster, M. and Campbell, K. (2005) An Evaluation of Partnership Development in the Construction Industry. *International Journal of Project Management*, 23(8), 611–621.

Bresnen, M. and Marshall, N. (2000) Partnering in construction: a critical review of issues, problems and dilemmas. *Construction Management and Economics*, 18(2), 229–237.

Cox, A. and Ireland, P. (2006) Relationship management theories and tools in project procurement. In: Pryke, S.D. and Smyth, H.J. (eds.) *Management of Complex Projects: a Relationship Approach.* Blackwell, Oxford, pp. 251–281.

Cox, A. Ireland, P. and Townsend, M. (2006) *Managing in Construction Supply Chains and Markets.* Thomas Telford, London.

Danermark, B., Ekström, M., Jakobsen, L. and Karlsson, J.C. (2002) *Explaining Society: Critical Realism in the Social Sciences.* Routledge, Abingdon.

Engwall, M. (1998) The project concept(s): on the unit of analysis in the study of project management. In: Lundin, R.A. and Midler, C. (eds.) *Projects as Arenas for Renewal and Learning Processes*. Kluwer Academic Publishers, Boston.
Gareis, R. (2004) Management of the project orientated company. In: Morris, P. W.G. and Pinto, J.K. (eds.) *The Wiley Guide to Managing Projects*. John Wiley & Sons, New York, 123–143.
Gilligan, C. (1982) *A Different Voice: Psychological Theory and Women's Development*. Harvard University Press, Boston.
Green, S.D. (2006) Discourse and fashion in supply chain management. In: Pryke, S.D. and Smyth, H.J. (eds.) *Management of Complex Projects: a Relationship Approach*. Blackwell, Oxford, 236–250.
Griseri, P. (2002) *Management Knowledge: a Critical View*. Palgrave Macmillan, Basingstoke.
Gummesson, E. (2002) *Total Relationship Marketing*. Butterworth-Heinemann, Oxford.
Handy, C.B. (1997) *Understanding Organizations*. Penguin, London.
Johnson, G., Scholes, K. and Whittington, R. (2005) *Exploring Corporate Strategy*. Financial Times-Prentice Hall, London.
Morris, P.W.G. (1994) *The Management of Projects*. Thomas Telford, London.
Morris, P.W.G. and Pinto, J.K. (eds.) (2004) *The Wiley Guide to Managing Projects*. John Wiley & Sons, New York.
Olayinka, R. and Smyth, H.J. (2007) Analysis of types of continuous improvement: Demonstration Projects of the Egan and post-Egan agenda. *Proceedings of RICS Cobra 2007*, 6–7 September, Georgia Institute of Technology, Atlanta.
PMI (2004) *A Guide to the Project Management Body of Knowledge (PMBOK® Guide)*. Project Management Institute, Newton Square.
Partington, D., Pellegrinelli, S. and Young, M. (2005) Attributes and levels of programme management competence: an interpretive study. *International Journal of Project Management*, 23(2), 87–95.
Pryke, S.D. (2004) Analysing construction project coalitions: explaining the application of social network analysis. *Construction Management and Economics*, 22(8), 787–797.
Pryke, S.D. (2005) Towards network theory and project governance. *Construction Management and Economics*, 23(9), 927–939.
Pryke, S.D. (2008 forthcoming) *Supply Chain Management in Construction*. Blackwell, Oxford.
Pryke, S.D. and Smyth, H.J. (2006) *The Management of Complex Projects*. Blackwell, Oxford.
Robbins, S.P. (2003) *Organisational Behaviour*, 10th Edition. Prentice Hall, New Jersey.
Ruddock and Knight (in press) *Advanced Research Methods in Construction*. Blackwell, Oxford.
Sayer, R.A. (1992) *Method in Social Science: A Realist Approach*, 2nd Edition. Routledge, London.
Smyth, H.J. (2000) *Marketing and Selling Construction Services*. Blackwell Science, Oxford.
Smyth, H.J. (2006) Competition. In: Lowe, D. and Leiringer, R. (eds.) *Commercial Management of Projects: Defining the Discipline*. Blackwell, Oxford, pp. 22–39.

Smyth, H.J. (2011 forthcoming) *Core Competencies in Construction.* Blackwell, Oxford.

Smyth, H.J. and Edkins, A.J. (2007) Relationship management in the management of PFI/PPP projects in the UK. *International Journal of Project Management*, 25(3), 232–240.

Smyth, H.J. and Fitch, T. (2007) Relationship management: a case study of key account management in a large contractor. *Proceedings of CME25: Construction Management and Economics: Past, Present and Future*, 16–18 July 2007, University of Reading.

Smyth, H.J. and Morris, P.W.G. (2007) An epistemological evaluation of research into projects and their management: methodological issues. *International Journal of Project Management*, 25(4), 423–436.

Smyth, H.J., Morris, P.W.G. and Kelsey, J.M. (2007) Critical realism and the management of projects: epistemology for understanding value creation in the face of uncertainty. *Paper presented at Euram 2007*, May 16–19, HEC and INSEAD, Paris.

Thiry, M. (2004) Program management: a strategic decision management process. In: Morris, P.W.G. and Pinto, J.K. (eds.) *The Wiley Guide to Managing Projects.* John Wiley & Sons, New York, 257–287.

Walker, A. (2007) *Project Management in Construction.* Blackwell, Oxford.

Wenger, E. (1998) *Communities of Practice: Learning, Meaning and Identity.* Cambridge University Press, Cambridge.

Williamson, O.E. (1979) Transaction-cost economics: the governance of contractual relations. *Journal of Law and Economics*, 22(2), 233–261.

Winch, G.M. (2002) *Managing the Construction Project.* Blackwell, Oxford.

Section I

Collaborative Relationships in Contractual Frameworks

This section concerns collaborative relationships in contractual frameworks. It explores the practical scope and conceptual limits of collaboration within frameworks and programmes. Initiatives for continuous improvement based upon relational contracting are considered and the potential for further and future development is addressed. The chapters presented in this section feed into other sections and thus relate to other chapters too.

In **Chapter 1** Mason addresses specialist contractors and partnering. **Chapter 2**, by Potts, analyses some of the sweeping changes in the quantity surveying profession through the case study of Terminal 5, London Heathrow Airport. Lee and Egbu examine client project requirements and project team knowledge for refurbishment projects in **Chapter 3**. Finally, **Chapter 4**, by Kumaraswamy, Anvuur and Mahesh, explores contractual frameworks and cooperative relationships.

1 Specialist contractors and partnering

Jim Mason

Introduction and context

The introduction of partnering to the United Kingdom's construction industry represented a sustained effort to overcome its perceived performance problems (Barlow and Jashapara, 1998). Although earlier references to partnering exist, the means of introduction is generally acknowledged as starting with the recommendations made in Sir Michael Latham's Report *Constructing the Team* and progressing through the Latham-endorsed Construction Task Force Report *Rethinking Construction* (Egan, 1998). These documents have been described as an 'impetus for change' and the latter 'a framework for radical improvement and modernisation' (Wood, 2005).

Analyses of the theory behind partnering, the means of its introduction and its achievements to date at the employer–main contractor level have already been the subject of many other papers and textbooks (e.g. Bennett and Jayes, 1998). The starting point for this chapter is that partnering has made a substantial impact at certain levels of the industry. Whether this amounts to a fundamental shift in how business is conducted, as claimed by some commentators (Beach *et al.*, 2005), is debatable. However, partnering represents perhaps the most significant development to date as a means of improving project performance whilst offering direct benefit to clients and contractors (Wood, 2005).

Regardless of how fundamental the shift in thinking, the success of a concept is measured by its take-up. In the case of partnering the take-up along the supply chain is currently small (Olayinka and Smyth, 2007). The construction industry has a reputation for adapting slowly to change especially at the specialist contractor level. Competitive tendering remains the principal mechanism for sub-contractor selection particularly for non-specialist services (Dainty *et al.*, 2001). It has also been accurately observed that strategic partnering alliances, generally referred to as partnering, are infrequent in the construction industry (Shimizu and Cardozo, 2002).

Perhaps the expectation of a higher incidence of partnering is premature. It has been observed that the process of change is in its early stages (Wood, 2005) and that whilst the Latham and Egan Reports represent current aspirations for the future direction of the construction industry, their effects both in terms of management and legal terms, remain to be established (Uff, 2005; Pryke, 2006).

Another potential explanation for the slow take-up of partnering is confusion surrounding its definition. Partnering has been described as a generic term for a variety of formal and less formal arrangements (Beach, 2005), with at least half a dozen different perspectives on partnering (Matthews *et al.*, 2000).

This last point is probably slightly wide of the mark in that a consensus exists as to the essential ingredients of successful partnering. Essentially the relationship is based on trust, dedication to common goals and an understanding of expectations and values (Matthews *et al.*, 2000). However, the cause of partnering would be helped if construction as an industry articulated an agreed philosophy for partnering and identified the appropriate benchmarks (Hibberd, 2004). The work of Constructing Excellence in recent years has improved the position here, but issues still remain as to whether their message is being heard at the specialist contractor level.

Where a definition is elusive it is often easier to identify what something is not rather than what it is. Partnering is the antithesis of open market competitive tendering – a process with inherent tensions and conflicts between clients and suppliers driven in different directions due to the nature of the competitive environment (Barlow and Jashapara, 1998). Relationships based mainly on lowest price run the risk of being distrustful if not antagonistic, and rooted in the fear that the other party might engage in opportunistic behaviour (Beach *et al.*, 2005). Specialist contractors are particularly vulnerable in the competitive environment because main contractors realise the greatest potential for cost savings lies with sub-contractors (Matthews *et al.*, 2000).

It would be clearly wrong to portray all opinions on competitive tendering as negative. Equally, not all opinions on partnering are positive and it would be equally wrong to accept the claims made at face value. The demonstration projects selected by the Strategic Forum for Construction have consistently exceeded Egan's targets (Beach *et al.*, 2005) and a study of 291 construction projects showed a positive relationship between partnering activities and project success (Larson, 1997). However, not everyone is convinced by the claims. There is little critical analysis of sufficient empirical depth to be convincing (Wood, 2005) and the research is notable for its heavy reliance on anecdotal evidence concentrating on 'exemplar' organisations (Bresnen and Marshall, 2000a).

Neither is partnering without its detractors. There is concern that partnering prevents new companies from entering closed markets and reduces potential business opportunities (Davey *et al.*, 2001). There is also concern that practical constraints need to be overcome, including difficulties in providing continuity of work, and there are misgivings about long-term relationships being too 'cosy' and uncompetitive (Bresnen and Marshall, 2000c). Opinions about what might happen to partnering if there is a downturn in demand are also more cautious (Wood, 2005).

Specialist contractors are of vital importance to the construction industry and their contribution to the total construction process can account for as much as 90% of the total project spend (Nobbs, 1993). For the average specialist, issues of survival and continuity of work still dominate their decision-making process and unless partnering can convince these firms that it can improve their chances it is unlikely to have a significant impact (Packham *et al.*, 2001). The context of this study is probably best summed up in the statements that whilst attitudes towards sub-contractors have improved over the past 20 years, they have not improved nearly enough (Love, 1997).

Aims and objectives

The literature demonstrates that the position of specialist contractors in relation to partnering is a complicated one. Competitive tendering has been the norm for so long that there is a suspicion of any new initiatives, particularly those dictated from 'upstream' in the supply chain. If the literature is correct then only a minority of specialists is aware of partnering and only a smaller number again will have had partnering experiences. Amongst those with knowledge and experience there may well have been economic and cultural factors weighing against the likelihood of a successful and positive experience including the continuing opportunistic behaviour of main contractors.

However, there is a lack of empirical qualitative research in this important field capable of testing these notions. The aim of this chapter is therefore to inform the debate about the impact partnering had made on specialist contractors and their views and experiences. From this aim a number of objectives for the research underpinning this chapter can be identified:

1. Determine the level of knowledge about partnering amongst specialist contractors.
2. Identify the range of experiences and practices being adopted in current partnering arrangements.
3. Assess the actual and potential barriers to success and the benefits accruing to specialist contractors through partnering.
4. Consider whether real change is being effected through partnering and whether such change might continue.

Research methodology

The existence of deep-rooted opinions amongst the tiers of the supply chain about each other's position and performance is accepted as fact. The perspective put forward by this chapter is that these deeply rooted opinions are overly simplistic and ignore important messages that are available through quantitative and qualitative study of the views and experiences of those involved. The opinion that specialist contractors are unwilling and/or unable to participate in partnering-type arrangements, formed the proposition for this research.

This proposition is tested through the collection of primary data, firstly in the form of questionnaires and subsequently through semi-structured interviews. The questionnaire was designed to capture the views and experiences of senior individuals involved in tendering and winning work for specialist contractors. The respondents were asked to respond to the same set of questions in a predetermined order (Gray, 2004). The results of the questionnaire were analysed in order to detect common themes, issues, opinions and the degree of consensus or otherwise amongst the body of respondents.

Upon studying these findings a number of follow-up questions were identified. Adopting a flexible approach, the natural progression for the research to take was to deliver the follow-up questions in semi-structured interviews with the original questionnaire respondents. This qualitative methodology provides a data richness which the questionnaires alone would not have captured. Direct quotations from the follow-up interviews are used extensively in the presentation of the findings.

The sample selected seeks to give the study credibility by providing a sufficient number and range of experiences and views of partnering amongst specialist contractors. The Confederation of Construction Specialists is an organisation with some 300 plus members which was set up to achieve real and beneficial improvements in the business environment in which specialist contractors operate. The Confederation was approached to assist in the data collection because of its considerable size and variety in the profiles of its member organisations. The Confederation allowed its members to be approached and 30 firms returned completed questionnaires during 2005–2006. Upon submitting their questionnaires these firms where invited to take part in a semi-structured interview. Ten of these interviews were conducted during the first 6 months of 2006.

The questionnaire contained a variety of closed questions, rating scales and 'forced choices' allowing for a variety of individual responses. The profile of the respondents is demonstrated in Figures 1.1 and 1.2. The experiences of the specialists of partnering captured by the questionnaire are presented in Figure 1.3. The views of the specialists are presented in Figure 1.4.

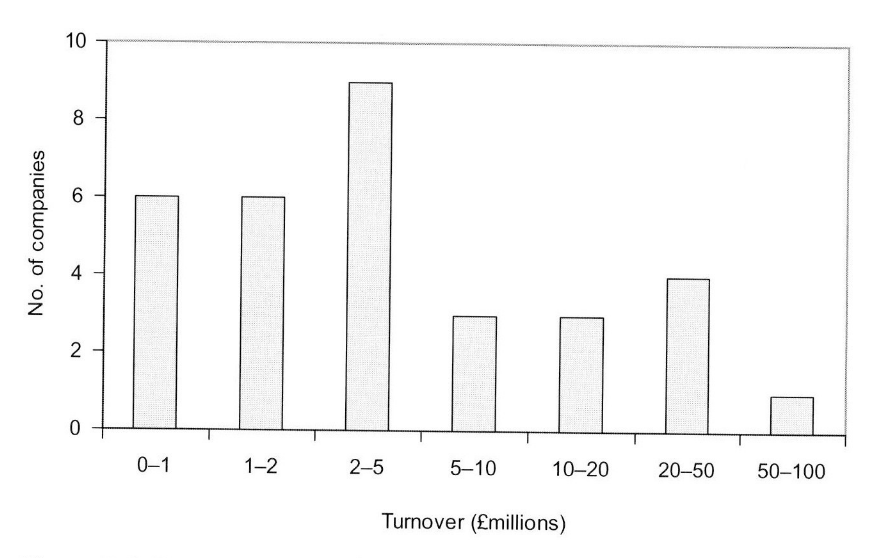

Figure 1.1 Questionnaire respondents by turnover

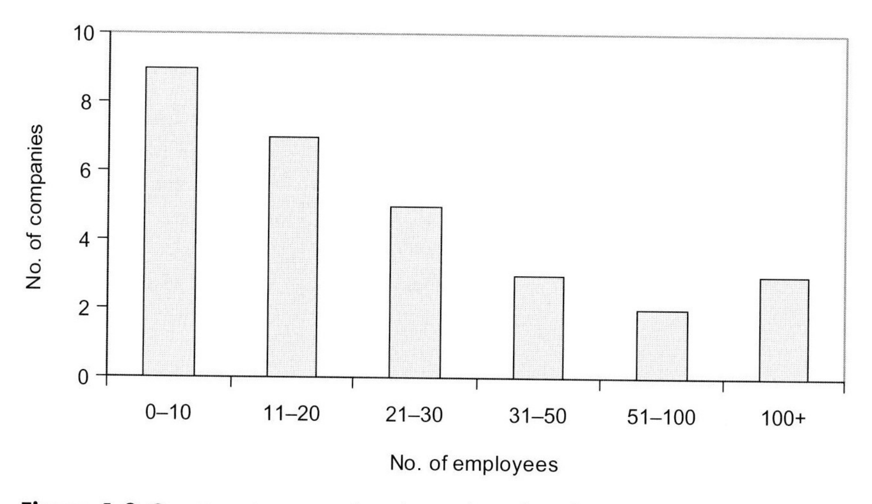

Figure 1.2 Questionnaire respondents by number of employees

Findings

As demonstrated by Figure 1.3, there was a high incidence of experience and familiarity with partnering. These statistics are extremely positive in terms of the impact of partnering even allowing for the self-selecting nature of the specialists responding to the questionnaire. Figure 1.4 also has some positive findings for partnering – most specialists agree on

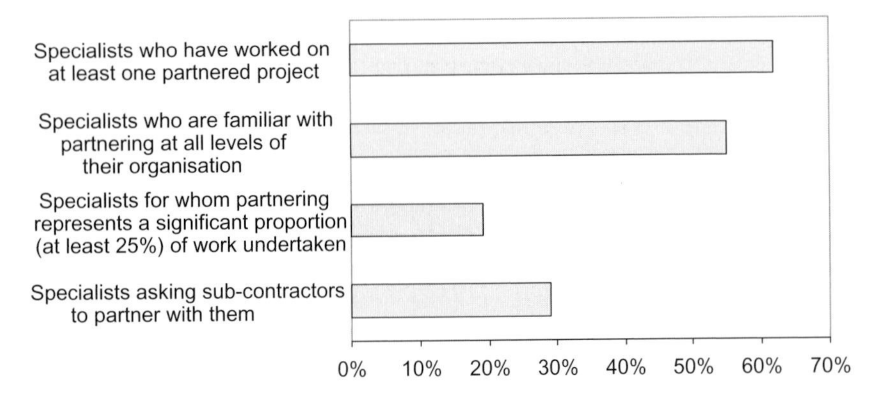

Figure 1.3 Partnering experiences of specialists

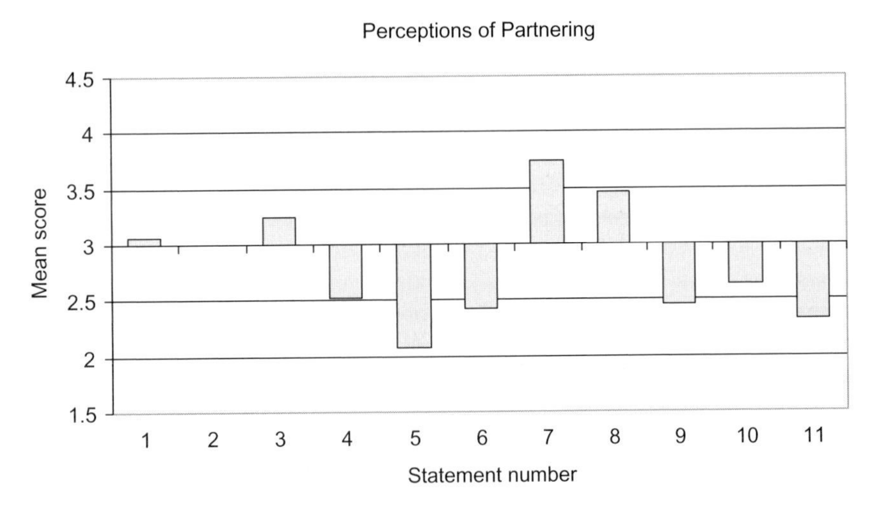

1 = Strongly agree 2 = Agree on balance 3 = No strong feelings 4 = Disagree on balance
5 = Strongly disagree for the perceptions of partnering from the following schedule of statements:

1 Longer-term arrangements with contractors/clients the norm
2 Team working between organisations has improved
3 The number of conflicts on projects has reduced
4 Partnering benefits only felt at main contractor/employer level
5 Sub-contractors stand to gain by properly partnered work
6 Partnering will grow in importance
7 Partnering has restricted competition – bad for all concerned
8 Partnering is hot air with few if any actual benefits
9 On the whole partnering is a positive development
10 It is too early to say what effects have been/will be
11 Partnering is best achieved informally

Figure 1.4 Mean score

balance that specialist contractors stand to gain from partnering which they see as a positive development that will grow in importance in the coming years.

The respondents were less positive about the state of the construction industry with 58% of the view that conflict levels had at least stayed the same in recent years and the same percentage holding the opinion that team working had not improved. Another interesting finding was that of the 81% of respondents familiar with partnering, the majority (65%) claimed familiarity at all levels of their organisations.

The 'other comments' section of the questionnaire gave valuable insight into some of the commonly held views and experiences of partnering. These views directly influenced some of the questions in the follow-up semi-structured interviews.

The interview questions sought to probe further into some of the views expressed in the questionnaire stage of the research. The interviews were conducted over the telephone and were recorded for subsequent analysis. The interviewees were asked in more detail and were able to expand on:

- How work was won – the incidence of partnering.
- Experiences and views on partnering.
- Company's level of familiarity with partnering.
- Definitions of partnering.
- Whether the company partnered with suppliers/sub-contractors.
- How they perceived the state of the industry.
- Views on the future of partnering.
- Comments on barriers to partnering.
- Overall views.

The presentation of the findings groups together in themes of views expressed by the specialists approached. Some particular views appear in direct quotations whereas widely held views are presented in the narrative itself.

The existing situation

As highlighted in the literature review, competitive tendering remains dominant. The specialists were frequently one of four or five approached on an approved list and work was awarded to the successful applicant on the basis of lowest price tendering.

Where competitive tendering is used there was also a high incidence of imposition of main contractor's own terms and conditions on the specialists. Where specialists did have their own terms and conditions, these were often excluded.

Some specialists were required to enter two-stage tendering procedures possibly in line with the main contractor's own tender requirements to the client. For one specialist it was only at the relatively late stage of progressing to the second stage of the tender process that the first mention of partnering on the project was raised.

Partnering had made a favourable impression with some of the specialists approached expressing a preference for it. Others remarked that some clients, particularly local authorities, now require them to partner directly without necessarily any main contractor involvement:

We have used partnering charters and attended partnering meetings. We had one fruitful experience where we achieved minor miracles on a very complicated job which was done in quick time.

However, the predominant view held by specialists is that partnering has not improved relationships or the level of disputes experienced by the industry. Pessimism abounded in this area amongst specialists, with one individual claiming that the 'same old battles and excuses' for withholding payment were being experienced.

Perceptions of partnering

In analysing the comments, attempts were made to distinguish between those specialists with direct experience of partnering and those specialists relying on received information outside of their own experiences. This distinction is important in order to assess whether the views held reflect direct experience or otherwise. In attempting the distinction those with a general view are separated from the particularly held views. Of the two there was a greater frequency of cynical views amongst the written comments of generally stated views.

The generally stated view was that partnering does not extend down the supply chain, that specialists are usually deliberately excluded from it and that where they are included it benefits only the main contractor and the client. There were instances where these views were backed up by those with direct experience such as describing a partnering charter as 'pie in the sky' and partnering as 'just a fancy name'. Others felt uncomfortable with partnering arrangements where 'nothing was put in writing'.

The view amongst most of those with direct experience was a positive one. One specialist stated partnering was a good idea for like-minded contractors and specialists to work together on a regular basis. In the words of another specialist with direct experience:

We have been using frameworks based on 3–4 years work – all problems were ironed out at the beginning with no hidden surprises and good team work.

The level of knowledge

The follow-up questions revealed a less convincing response about how far knowledge had spread than the questionnaires had indicated could be expected. Where familiarity with partnering did exist it was limited to senior management and even here knowledge was described as patchy and insufficient:

> *The management are insufficiently familiar with partnering – they know the theory.*

The definitions of partnering put forward by the specialists were more encouraging in terms of indicating a good grasp of the subject. Definitions ranged from 'respecting each other's reasonable aspirations' to 'becoming involved at all stages' and 'having a relationship without stand up fights over payments'.

Perhaps slightly more worryingly, no mention of the competitive advantage available through partnering was made by the specialists. All the emphasis was on the relationship side of partnering with one specialist going as far as describing the concept as 'acting as one big family'.

Positive indicators

A high incidence (28% of questionnaire respondents) was recorded of specialist contractors partnering with their own sub-contractors and suppliers, even without upstream partnering being specified. A significant number of the respondents recognised that partnering with their supply chain was highly desirable for everyone concerned:

> *Even on a non-partnered job we look to use partnering techniques downstream.*

Another specialist pointed out that the benefit for their supply chain was continuity of work, surety of payment, improved techniques which enabled improvement of output and ultimately improvement of firm's margins.

Another pointed out that they relied absolutely on their sub-contractors and that they were thinking about formalising the arrangement in some way. The length of the relationship with sub-contractors, which one specialist put at between 10 and 20 years, was also seen as a key driver for partnering with sub-contractors.

Barriers to success

There were some interesting indicators of the pre-conditions necessary to improve the working of partnering amongst those specialists with particular views based on their experience. The role of the client was seen as pivotal with the view expressed that the client must be committed and intelligent with the stamina to see through hard times as well as good ones. The virtues required for successful partnering were also touched upon by some specialists – willingness, capability and honesty:

Partnering is best when client led – it has a better feel to it.

As one might expect, there was some berating of main contractor's behaviour amongst the specialists such as the inappropriate allocation of risk where main contractors could not be 'bothered to deal with it':

Partnering for the main contractor is customer focused and does not include the sub-contractors.

Complaints about main contractor practice were not, however, across the board and some main contractors were praised for 'becoming more professional'. One specialist spoke of partnering *champions* within main contractor organisations who had commitment and integrity to the concept of partnering. The same specialist recognised that others within the same organisation might not have had the same approach. This leads into the comment of another specialist that the personnel in main contractors and clients were not static for long enough for enduring relationships to be created.

The future of partnering

Views on the long-term prognosis for partnering were mainly bleak, with some positive exceptions. Again, the distinction between those with particular views based on experience and those with a general view is an interesting one.

The specialists with particular views were more positive about the future of partnering. One specialist in contact with housing associations foresaw a greater incidence of partnering in 10 years based on how he saw clients procure work. Another specialist thought that savings of 30% would be possible through the continued implementation of partnering.

Those without first-hand experience or limited to partnering through main contactors held mainly negative views. Ten years was not thought of as long enough by one specialist to bring about the cultural change required. Another dismissed partnering as too bureaucratic a process and merely as a 'management tool for those who want to use it'. Education was seen as key by another specialist in bringing about change:

You need to give something more than a fancy title – culture will take for ever to change when you can get away with things so readily.

Main discussion

Competitive tendering obviously remains as the principal method for sub-contractor selection. These interviews concur with the finding that a few contractors are experimenting with sub-contractor partnering, while for the majority it is business as usual (Greenwood, 2001). The first point for discussion is that the impact of partnering is such that the specialists approached had an opinion on it without necessarily having personal experience. In the absence of personal experience specialists tend to rely on the views of their peers and leading figures in the construction industry press. World-weary cynicism appears more contagious than ringing endorsement and the image of partnering seems to have suffered accordingly. This was apparent in the responses to the question on the future of partnering where a pessimistic picture was painted. Bridging this 'knowledge gap' from received wisdoms to positive first-hand experiences of partnering is one of the challenges the concept faces.

The definitions of partnering elicited from the interviewees are consistent with the elements identified as trust, dedication to common goals and an understanding of each other's individual expectations and values (Matthews *et al.*, 2000). These definitions of partnering are relatively simple and straightforward. However, the definitions offered by the respondents lack one vital ingredient – competitive advantage. The bottom line is that the effectiveness of partnering comes down to what is in it for the partner and that often means money. There must be a business case for partnering otherwise it is unlikely to succeed. For partnering to be successfully adopted at the specialist contractor level the concept needs to be promoted on competitive advantage rather than the laudable but simple aims currently associated with partnering.

On the other hand, the strength of the simple central message is demonstrated by the incidence of the specialists recognising that they were partnering with their own sub-contractors. This in turn raises the question in the minds of the specialist as to why they themselves are not being partnered with on a more frequent basis. It is at this stage that the main contractor is usually vilified and identified as the major barrier to a higher incidence of successful partnering.

The front-line managers of main contractors have been criticised elsewhere by sub-contractors as inhibiting better integration and acting aggressively and preventing sub-contractor early involvement in projects (Dainty *et al.*, 2001). The same sub-contractors saw partnering-related practices, such as open-book accounting, merely as mechanisms for main contractors to drive down sub-contractors' profits.

To vilify the main contractor in this manner misses the point of the vital role of the client. A number of the interviewees expressed the view that their involvement was crucial. Clients need to take a much more participative role in teambuilding and the unwillingness of the client to commit to the partnering agreement has been seen as the main reason for ineffective project partnering (Barlow and Jashapara, 1998). Taking this point further, it is this chapter's submission that greater client involvement in partnering would in turn lead to greater specialist involvement. If a client (or their representative) never asks to see or engage with its partners then the main contractor can hardly be blamed if they continue to deal with their sub-contractors in their default manner based on competition and leverage of the supply chain. Improvements here require back-to-back relationships and conditions of contract between the client/main contractor/specialist contractor.

It has been noted elsewhere that contractors drawn in reluctantly at first to partnering become more questioning organisations after their experiences help them to acquire innovative techniques and new ideas by sharing information (Barlow and Jashapara, 1998). The specialist contractors interviewed for this chapter demonstrated similar characteristics and in every case would probably admit to having learned something positive from their partnering experiences.

However, at the same time the interviewees were unable to detect any discernible improvement in the state of the industry in terms of disputes and the incidence of longer-term working arrangements.

It may be the case that it is still too early in the life of partnering to expect major change or that the changes being made are more subtle ones than are readily discernible at the 'business end' of the industry where the specialists find themselves.

A final point that cannot be overlooked is the difficulty of partnering with specialist contractors who are only on site for a short time and possibly not heard from again from project to project. This raises the question for further study of the degree to which partnering is universally applicable to construction projects and its personnel. This would also involve a consideration of the different forms of partnering and their applicability to specialist contractors.

Conclusions and recommendations

Partnering is no longer new to the United Kingdom's construction industry. Most of the studies to date have been limited to examining the theory of partnering rather than its impact. The literature to date suggests that specialist contractors hold a negative view of partnering with a limited take-up due in part to an exclusion from participation.

Partnering has pierced the collective consciousness of specialist contractors. For those without experience of partnering the view held is usually negative with elements of cynicism. Amongst those with first-hand knowledge of partnering the majority view is positive, although heavily qualified by past experiences and mistrust of main contractors.

The proposition that specialist contractors are unwilling and/or unable to enter into partnering arrangements is rejected. Specialists can, in the right conditions, contribute meaningfully and prosper in the collaborative working environment. The incidence of the right conditions is relatively low and is further handicapped by the poor perception of partnering in practice amongst the specialists themselves.

To improve this perception of partnering amongst specialist contractors this chapter makes the following recommendations:

- A re-examination of the client's role in partnering contracts. If the client never asks to see its partners then the wrong messages are being sent to the specialists involved. The client and the specialist contractors should be brought closer together for the benefit of everyone concerned.
- Promotion of partnering on the grounds of adding competitive advantage for specialist contractors rather than on the grounds of mutual trust and cooperation. The experiences of specialist contractors leave them unmoved by the more laudable aims of partnering and the emphasis needs therefore to be shifted.
- Focusing attention on successful examples of collaborative working such as the use of zero retentions and risk registers.
- Current informal downstream partnering involving specialist contractors has the potential to be encouraged and/or formally developed, hence cascading collaborative practice down the supply chain.

References

Barlow, J. and Jashapara, A. (1998) Organisational learning and inter-firm partnering in the UK construction industry. *The Learning Organisation*, 15(2), 86–98.

Beach, R., Webster, M. and Campbell, K. (2005) An evaluation of partnership development in the construction industry. *International Journal of Project Management*, 23, 611–621.

Bennett, J. and Jayes, S. (1995) *Trusting the Team: The Best Practice Guide to Partnering in Construction. Reading.* Centre for Strategic Studies in Construction/ Reading Construction Forum, Reading.

Bennett, J. and Jayes, S. (1998) *The Seven Pillars of Partnering.* Thomas Telford, London.

Bresnen, M. and Marshall, N. (2000a) Partnering in construction: a critical review of issues, problems and dilemmas. *Construction Management and Economics*, 18, 229–237.

Bresnen, M. and Marshall, N. (2000b) Motivation, commitment and the use of incentives in partnerships and alliances. *Construction Management and Economics*, 18, 819–832.

Bresnen, M. and Marshall, N. (2000c) Building partnerships: case studies of client–contractor collaboration in the UK construction industry. *Construction Management and Economics*, 18, 819–832.

Bresnen, M. and Marshall, N. (2002) The engineering or evolution of co-operation? A tale of two partnering projects. *Construction Management and Economics*, 20, 497–505.

Dainty, A., Briscoe, G. and Millett, S. (2001) Subcontractor perspectives on supply chain alliances. *Construction Management and Economics*, 19, 841–848.

Davey, C., Lowe, D. and Duff, A. (2001) Generating opportunities for SMEs to develop partnerships and improve performance. *Building Research and Information*, 29(1), 1–11.

Egan, Sir John (1998) *Rethinking Construction.* Department of the Environment, Transport and the Regions, London. (http://www.rethinkingconstruction.org/)

Gray, A. (2004) *Research Practice for Cultural Studies: Ethnographic Methods and Lived Culture.* Sage, London.

Greenwood, D. (2001) Sub-contract procurement: are relationships changing? *Construction Management and Economics*, 19(1), 5–7.

Hibberd, P. (2004) The place of standard form of building contracts in the 21st Century. *Society of Construction Law Conference Paper*, 11 March 2004, www.scl.org.uk.

Larson, E. (1997) Partnering on construction projects: a study of the relationship between partnering activities and project success. *IEEE Transactions on Engineering Management*, 44(2), 188–195.

Latham, Sir Michael (1994) *Constructing the Team: Joint Review of Procurement and Contractual Arrangements in the United Kingdom Construction Industry.* HMSO, London.

Love, S. (1997) Subcontractor partnering: I'll believe it when I see it. *Journal of Management in Engineering*, September/October, 29–31.

Matthews, J., Leah, P., Phua, F. and Rowlinson, S. (2000) Quality relationships: partnering in the construction supply chain. *International Journal of Quality and Reliability Management*, 17(4–5), 493–510.

Ng, S., Rose, T., Mak, M. and Chen, S. (2002) Problematic issues associated with project partnering – the contractor perspective. *International Journal of Project Management*, 20, 437–449.

Nobbs, H. (1993) *Future Roles of Construction Specialists.* The Business Round Table, London.

Olayinka, R. and Smyth, H.J. (2007) Analysis of types of continuous improvement: demonstration projects of the Egan and post-Egan agenda. *Proceedings of RICS Cobra 2007*, 6–7 September, Georgia Institute of Technology, Atlanta.

Packham, G., Thomas, B. and Miller, C. (2001) Partnering in the Welsh construction industry: a subcontracting perspective. *Construction Innovation*, 5(1), 13–26.

Pryke, S.D. (2004) Twenty-first century procurement strategies: analysing networks of inter-firm relationships. *RICS Paper Series 4(27)*, RICS, London.

Pryke, S.D. (2006) Legal issues associated with emergent actor roles in innovative U.K. procurement: prime contracting case study. *ASCE Journal of Professional Issues in Engineering Education and Practice*, 132(1), 67–76.

Shimizu, J. and Cardoso, F. (2002) Subcontracting and cooperation network in building construction: a literature review. *Proceedings IGLC-10*, 6–8 August, Gramado, Brazil.

Uff, J. (2005) *Construction Law*, 9th Edition. Thomson Sweet and Maxwell, London.

Walliman, N. (2005) *Your Research Project*, 2nd Edition. Sage Publications, London.

Wengraf, T. (2001) *Qualitative Research Interviewing*. SAGE Publications, London.

Wood, G. (2005) Partnering practice in the relationship between clients and main contractors. *RICS Paper Series 5(2)*, RICS, London.

Appendix 1

	Trade	Turnover (£million)	Projects	Public/private	Employees
1	Flooring	0.8	50	50/50	18
2	Systems	21.5	450	80/20	360
3	Refurbishment	2	35	50/50	40
4	Concrete	1.25	8	33/66	30
5	Painting	1.5	4	90/10	20
6	Ventilation	20	250	100/0	160
7	Industrial doors	32	3	0/100	130
8	Ventilation	1.8	30	50/50	26
9	Fire alarms	16.5	1500	20/80	204
10	Power systems	4	N/A	0/100	27
11	Cutting/drilling	0.2	4	10/90	5
12	Industrial doors	4.5	20	10/90	64
13	Windows	4.75	27	40/60	13
14	Mechanical services	4.5	12	75/25	48
15	Road marking	5	25	96/4	72
16	Refrigeration	1	4	5/95	4
17	Foundations	100	100	10/90	850
18	Industrial doors	3	40	30/70	44
19	Metal doors	5	5	0/100	60
20	Cladding	10	13	0/100	150
21	M&E QS	6.3	60	20/80	100
22	Curtain walling	40	6	0/100	250
23	Curtain walling	5	12	40/60	36
24	Refrigeration	0.75–1	3	0/100	3
25	Concrete	12	15	0/100	160
26	Metalwork	2	24	75/25	40
27	Telecoms	0.2	4–5	50/50	4
28	Highways	6	15–20	90/10	70
29	Specialist doors	5	20	20/80	50
30	Concrete	22	25	50/50	100

2 Change in the quantity surveying profession

Heathrow Terminal 5 case study

Keith Potts

Introduction

This chapter deals with the efforts of one of the largest UK construction clients in making the transition from adversarial procurement and management practices to collaborative strategies. The chapter uses Heathrow Terminal 5 as a vehicle for discussion of these new collaborative relationships, particularly focusing upon the implications for professional service providers.

In the last decade procurement and project practice on both public and private works within the construction industry have been the subject of radical change. Numerous reports have identified both the public and private sector clients' dissatisfaction with the traditional approach within the industry. In the UK this change was initially driven by the Latham Report (1994), and the CIB Reports (1996/97) which identified public and private sector clients' dissatisfaction with the traditional approach within the industry.

Significantly Sir John Egan, then Chief Executive of BAA plc, identified that the industry as a whole was under-achieving and challenged all in the industry to fundamentally change its culture and methods of working, hence, *Rethinking Construction* (Egan, 1998). The Strategic Forum for Construction's Report *Accelerating Change* (2002) challenged the industry to maximise value for all clients, end-users and stakeholders and exceed their expectations through consistent delivery of world class products.

These reports were reinforced by the Achieving Excellence in Construction Procurement Guides from the Office of Government Commerce (2003) and the National Audit Office reports *Modernising Construction* (2001) and *Improving Public Services through Better Construction* (NAO 2005a and 2005b).

In his foreword to *Improving Public Services through Better Construction* (NAO, 2005a) Sir Michael Latham states:

> *Best practice is about partnering, collaborative working and stripping out of the equation at the earliest possible stage those costs which add no value. To achieve that, it is vital to include the whole supply chain.*

The Case Studies (NAO, 2005b) demonstrate that the construction industry is changing and that best value is being secured on public projects at a national level, for example, in the NHS through Procure 21. In the private sector, the Honda car company had achieved a 40% improvement in its UK construction performance over 11 years (as measured by the building cost for buildings of equivalent functionality) (Bayfield and Roberts, 2004).

At local government level best practice is being implemented through strategic alliances and partnering arrangements. A year into the first construction collaboration of its kind in the UK the Birmingham Construction Partnership has demonstrated 52% improvement in projects delivered to time and 29% improvement in projects delivered to budget (www.wmcoe.gov.uk/download.asp?id=450). Furthermore several enlightened local authorities are moving towards the use of the collaborative PPC2000 form of contract, for example, Stoke-on-Trent and Wolverhampton.

During this dynamic period of change quantity surveyors (QSs) have begun to reinvent themselves *moving up the chain of responsibility* with many becoming the client's lead consultant. Ashworth and Hogg (2000), when reviewing the development of quantity surveying, identified a shift in emphasis from *cost to value* with many QSs *extending the range of services* that they offer clients. At first this meant developing project management, followed by the provision of development appraisals, life-cycle costing, facilities management and other services. Furthermore, under partnering contracts there has been a significant shift in emphasis from claims management to proactive value and risk management.

This shift of emphasis from cost to value requires moving from cost control to a budget to actively looking for best-value solutions to enable project delivery that exceeds the client's expectations in terms of both budget and time. Also, adding value means helping the client and the client's team to make informed management decisions through risk identification and management and trend analysis of project control information, such as earned value analysis and production key performance indicators (KPIs).

A comparison of the range of services offered by QS firms 50 years ago, 25 years ago and today shows the development of the profession and the growth of additional services offered to clients including building surveying, facilities management, health and safety supervision (Davies, 2006). Many QSs are now providing clients with strategic advice at a much higher level than was previously the case; often underpinned with an increased knowledge of sophisticated electronic information management systems (Davies, 2006).

Cartlidge (2002) neatly summed up the challenge for the profession with the observation:

> *. . . quantity surveyors must get inside the head of their clients, fully appreciate their business objectives, and find new ways in which to deliver value and conversely remove waste from the procurement and construction process.*

Heathrow Terminal 5

The BAA Heathrow Terminal 5 (T5) is one of Europe's largest and most complex construction projects. T5 was approved by the Secretary of State on 20 November 2001 after the longest public inquiry in British history (46 months) and when completed in March 2008 it will add 50% to the capacity of Heathrow and provide a spectacular gateway into London.

The £4.3 bn project includes not only a vast new terminal and satellite building but nine new tunnels, two river diversions and a spur road connecting to the M25; it is a multi-disciplinary project embracing civil engineering, mechanical services, electrical systems, communications and technology contractors with a peak monthly spend of over £80 million employing up to 8000 workers on site. The construction of T5 consists of 18 main projects divided into 140 sub-projects and 1500 'work packages' on a 260 ha site.

Phase 1 construction of T5 is programmed for 5 years and can be broken down into five key stages:

- *Site preparation and enabling works (July 2002–July 2003)* – preparing the site for major construction activity. The work included a significant amount of archaeological excavation, services diversions, levelling the site, removing sludge lagoons and constructing site roads, offices and logistics centres.
- *Groundworks (Nov 2002–Feb 2005)* – included the main earthworks, terminal basements, connecting substructures and drainage and rail tunnels.
- *Major structures (Nov 2003–Sept 2006)* – the main terminal building (concourse A), first satellite (concourse B), multi-storey carpark and ancillary structures.
- *Fit out (Feb 2005–Sept 2007)* – significant items of fit out including building services, the baggage system, a track transit people-mover system and specialist electronic systems.
- *Implementation of operational readiness (Oct 2007–Mar 2008)* – ensuring Phase 1 infrastructure is fully complete and that systems are tested,

staff trained and procedures in ready for operation in Spring 2008 (BAA T5 fact sheet, *The Key Stages of Terminal 5*).

Phase 2, which includes a second satellite and additional stands, will start after 2006 when the residual sewage sludge treatment site will be vacated. When completed in 2010, the two phases will enable Heathrow to handle an additional 30 million passengers per year.

Project management philosophy

The project management approach on T5 was developed based on the principles specified in *Constructing the Team* (Latham, 1994) and *Rethinking Construction* (Egan, 1998) but went further than any other major project. The history of the UK construction industry on large-scale projects suggested that had BAA followed a traditional approach T5 would end up opening 2 years late, cost 40% over budget with six fatalities (Riley, 2005); this was not an option for BAA.

Significantly BAA expected a high degree of design evolution throughout the project in order to embrace new technological solutions and changes in security, space requirements or facilities functionality. On such a complex project early freezing of the design solution was not realistic. BAA realised that they had to rethink the client's role and therefore decided to take the total risk of all contracts on the project. BAA introduced a system under which they actively managed the cause (the activities) through the use of *integrated teams* which displayed the behaviours associated with partnering (Figure 2.1).

This strategy was implemented through the use of the T5 Agreement under which the client takes on legal responsibility for the project's risk. In effect, BAA envisaged that all suppliers working on the project should operate as a *virtual company*. Executives were asked to lose their company allegiances and share their information and knowledge with colleagues in other professions. This approach created an environment in which all

---

Risk management

BAA's approach to risk management has been a key factor in keeping the project on budget and ahead of schedule. Terminal 5 is being constructed under the T5 Agreement which means BAA acts as the prime client and accepts most of the risk. With this burden removed from contractors and suppliers, it enables everyone working on T5 to:

- Focus on managing the cause of problems, not the effects if they happen
- Work on truly integrated teams in a successful, if uncertain environment
- Focus on proactively managing risk rather than devote energy to avoiding litigation

---

Figure 2.1 BAA's approach to risk management (BAA T5 fact sheet *Risk Management*)

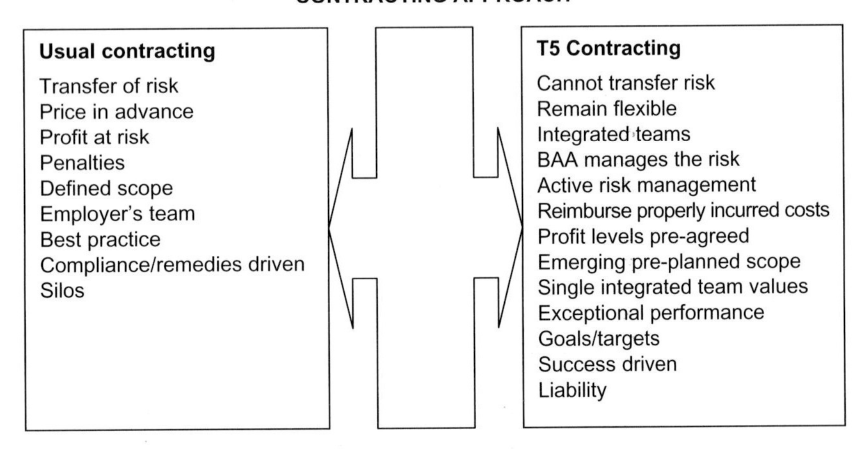

Figure 2.2 Traditional and Heathrow T5 contracting approach compared (BAA T5 factsheet *The Key Stages of Terminal 5*)

team members were *equal* and problem solving and innovation were encouraged in order to drive out all unnecessary costs, including claims and litigation, and drive up productivity levels (Douglas, 2005).

BAA's aim was to create one team, comprising BAA personnel and different partner businesses, working to a common set of objectives by the following means (Figure 2.2):

- The T5 Agreement with suppliers did not specify the work required; it is a commitment from the partner and a statement of capability, capacity and scope to be provided.
- The organisation was based on the delivery of products, seen as operational facilities, not a set of buildings.
- BAA selected the best people to suit the project's needs including 160 highly experienced and capable professionals from other organisations.
- By using collaborative software key information, such as the timetable, the risk reports, the work scope was freely available to the integrated project team.
- An Organisational Effectiveness Director with a team of 30 change managers provided training and support in order to implement the culture change required to work in an open and collaborative way (NAO, 2005b).

Many of the suppliers involved in T5 were brought on-board at the earliest stages of the planning process. This enabled completely integrated expert teams to work together to identify potential problems and issues before designs were finalised and fabrication and construction began. As a result the teams of suppliers and consultants were in a position to *add value* whilst designing safe solutions within the time, quality cost and safety targets. This approach encourages innovation, for example the development of pavement concrete has led to a 25% reduction in bulk materials required for the aircraft stands and pavement areas. This type of collaboration, whilst not unique in British construction, is still unusual.

T5 Agreement

The T5 Agreement is a unique legal contract in the construction industry. In essence it is a cost reimbursable form of contract in which suppliers' profits are ring-fenced and the client retains the risk. It focuses, in a collaborative style, on causes and management of risk through integrated team approaches. The reimbursable form of contract means there have been *no claims for additional payments and no payment disputes so far on the T5 project* (NAO, 2005a). This move away from a lump sum contract transfers a significant level of risk to the client and requires client-driven and -owned systems to manage risk and facilitate collaborative behaviour amongst the project actors.

BAA *uses cost information* from other projects, *validated independently*, to set cost targets. If the out-turn cost is lower than the target, the savings are shared with the relevant partners. This incentivises the teams to work together and innovate. It is an important step forward in improving profitability for constructors and specialist subcontractors: all other costs, including the profit margin, are dealt with on a transparent open-book basis (NAO, 2005b). BAA takes precautions against risk of the target being too high through a detailed 'bottom-up' analysis by independent consultants (Figure 2.3).

- It is a legally binding contract between Heathrow Airport Ltd and its key suppliers
- It stresses the importance of culture and behaviour are important. Innovatively, culture is specifically mentioned in the legal contract. The values – commitment, teamwork and trust – are key
- It addresses risk and reward. BAA holds the overall delivery risk. Suppliers take their share of the financial consequences of any risk to the project. And they share in the financial rewards of success (like the project finishing on time and within budget)
- Risk payments, which would normally be costed into a supplier's quote, have been held by BAA and drawn down if a risk occurred
- Key project risks have been insured – loss or damage to property, injury or death of people and, innovatively, professional indemnity for the project as a whole

Figure 2.3 Some key features of the Heathrow T5 Agreement (NAO, 2005b)

The T5 Agreement focuses on managing the cause and not the effect in order to ensure greater success levels in an uncertain environment. High performance levels and high benchmarking standards are demanded from all parties:

> *The idea is to have the best brains in all companies working out solutions to problems not working how best to defend their own corner.* (Comment by T5's Commercial Director Matthew Riley – within Broughton, 2004:36)

The T5 Agreement creates a considerable incentive for performance. If the work is done on time, a third of the savings goes to the contractor, a third goes back to BAA and a third goes into the project-wide pot that will only be paid at the end (Douglas, 2005). Any payment is dependent on meeting milestones set in that agreement. Suppliers also benefit from ring-fenced profit and an incentive scheme that rewards both early problem solving and exceptional performance.

The final strand to the T5 Agreement is the insurance policy. BAA has paid a single premium for the multi-billion project for the benefit of all suppliers, providing one insurance plan for the main project risk. The project-wide policy covers construction all risk and professional indemnity.

The T5 Agreement allows the project to adopt a more radical approach to the management of risk including early risk mitigation. Key messages include: 'working on T5 means everyone anticipating, managing and reducing the risks associated with what we're doing' (OGC, 2003).

Due to the inherent risks buried in the second- and third-tier contractors, BAA stipulate in the contract with the first-tier contractors their expectations on how second-tier suppliers are engaged. Whilst this has its inherent difficulties BAA felt that as they were carrying the risk they did not want key suppliers' contracts being let at fixed prices in high-risk areas if it presented a delivery risk to T5. BAA even went so far as to produce a draft second-tier supplier contract template, in some instances, in an effort to drive the T5 Agreement ethos down the supply chain (Ferroussat, 2005).

The legally binding contract centres on a 250-page handbook, containing the same set of conditions for each supplier (Table 2.1). Beneath that is a series of two to three page supporting documents defining particular capacities. These supporting documents can evolve as the working environment changes – flexibility is built in.

Controlling time, cost and quality

Keeping the project on schedule and within budget was obviously critical on a project of this size. Traditionally the two elements tend to run separately, often in two separate networks of activity: planning and costing.

Table 2.1 Heathrow T5 Documents: how they fit together (BAA T5 factsheet *The Key Stages of Terminal 5*)

Documents	Description
T5 Agreement	The terms and conditions everyone working on T5 is bound by
Supplemental Agreement	The detail of the agreement which is signed by the suppliers. It defines the work they're doing on T5
Functional Execution Plan	The support required to enable projects to deliver
Sub-project Execution Plan	The team's plan of work
Work Package Execution Plan	This is the breakdown of work by the supplier (combines preliminaries, specifications and drawings)
Support documents	
Commercial Policy	
Programme Handbook	
Core Processes and Procedures	
Industrial Relations Policy	

BAA, as client, aimed to be in the forefront of project control and on the T5 project was one of the first major users of the *Artemis* project management system in UK construction. The system is robust and can show how each area of the project is performing relative to target, on both schedule and costs. A further key point of the Artemis system is that it can give information at programme or at individual project level or sub-project level. Cost and performance data can be analysed in various ways, including the production of two highly useful indices, the Schedule Performance Index and the Cost Performance Index, which are generated for all the levels and for each package (Whitelaw, 2004). This enables trends to be identified, highlights where performance is not as planned and, most importantly, it enables informed management decisions to be made to keep the project on track.

The T5 project had a culture of 'right first time with no waste'. To implement this theme BAA had decided to pick up the costs if a contractor gets something wrong, arguing that that they would be much more likely to own up quickly to the mistake and hence save a great deal more money (and time) when the mistake came to light anyway. However, a contractor was not reimbursed for getting the same thing wrong twice; neither did it cover fraudulent behaviour by contractors or suppliers.

Quality management was dealt with through monthly review meetings for each of the 16 main projects and a monthly audit schedule workshop (Geoghegan, 2005).

The use of the NEC

Around 10% of the T5 value was procured under NEC contracts. The 70 first-tier suppliers were contracted under BAA's bespoke T5 Agreement.

Under this arrangement each first-tier supplier is responsible for developing their supply chain to deliver the work. BAA was recommending they use BAA's version of the NEC Engineering and Construction Contract for contracts with the thousands of second-tier suppliers – the only form recommended. This form was amended to work in line with the T5 Agreement and ensured that certain risks, i.e. insurance excess deductibles, were not passed down the supply chain.

BAA also used various NEC contracts, particularly the Professional Services Contract, for around 50 direct relationships with consultants and other suppliers.

Role of the cost consultants/contractor's quantity surveyors

BAA selected a consultancy framework for cost consultancy on the T5 project comprising a collaborative venture involving the Turner & Townsend Group and EC Harris Group Ltd (the collaborative vehicle known as TechT). Both companies were selected under the same terms of commission and each provided 50% of the staff. On this project these two major consultancy companies became one team 'joined at the hip'. At its peak the cost consultancy team comprised 120 staff, approximately two thirds of whom were QSs. The cost consultant was part of the integrated team providing commercial assurance for BAA and demonstrating value for money, for example through benchmarking. Working in an integrated environment meant developing trusting and collaborative working relationships with all suppliers whilst providing commercial assurance to BAA. Also, due to technological advances, time spent on the more mundane (but essential) activities was reduced which enabled a greater amount of time to be spent on value adding activities.

Laing O'Rourke Infrastructure Ltd was selected as the major first-tier supplier responsible for the civil construction, infrastructure and logistics delivery. Laing O'Rourke were involved in nearly 50 sub-projects with a turnover over the last 3 years averaging at £20 million per month managed by a team comprising more than 50 QSs (Simpkins, 2005). Significantly on this project BAA required the first-tier contractors and TechT to work as collaborative teams; there was no 'us and them' on T5.

Research

An in-depth literature search was undertaken in order to identify and better understand BAA's project management philosophy on T5. A case-study approach allowed the organisational, managerial, political and other dynamics influencing the project to be understood better so that the reality of such projects could be better appreciated (Morris and Hough, 1987).

The T5 project was selected as the basis for the case study examination for the following reasons:

- Sir John Egan, author of *Rethinking Construction* (Egan, 1998), was Chief Executive of BAA from 1990 until 1999.
- BAA are one of the champions of changing the culture and commercial environment in the industry.
- BAA have embraced the best practice principles of lean thinking and the use of integrated supply teams.
- BAA aim to be at the forefront in the development of effective project control systems.
- T5 was one of the largest projects in Europe and has harnessed some of the best brains in the industry.
- The lessons learned on the T5 project need to be understood and widely disseminated throughout the industry.

The literature search was followed by development of a questionnaire seeking to identify the roles of the cost consultants and the contractor's QSs and the techniques and tools used by them on the project. The questionnaire was developed based on a template which reflected the competencies identified within the Royal Institution of Chartered Surveyors (RICS) Assessment of Professional Competence (APC) Requirements and Competencies for construction surveyors who may be working as a consultant or for a contracting or an engineering company (RICS, 2002).

This approach was based on the fact that the RICS, as the main professional body representing QSs, should have expert knowledge of range of duties executed by their members. The only weakness of this approach might be if the list of competencies might not be comprehensive or current. However on balance the RICS APC list of competencies describing the duties of the QS was considered the most comprehensive and definitive list available. The methodology used was based on the tradition of empiricism, that is, based on observation not theory, and limited to one case study.

The main method used in this research was development of a questionnaire with follow-up interviews. The questionnaire was sent to representatives of the main players and was reinforced by email contact and telephone interviews with three senior representatives of the key players on the T5 project. These were EC Harris, Turner & Townsend and Laing O'Rourke. The main strength of this approach was that these participants had a depth of experience and a real understanding of the key issues involved. However it is accepted that additional contributions from representatives of other parties, for example, BAA,

Construction Manager MACE and major suppliers, would have been beneficial.

Research findings

BAA's cost consultants, initially 'T&T and ECH' and then 'TechT', and Laing O'Rourke's QSs provided both strategic and delivery services from inception through to the construction phase and contributed to:

1. Preparing development appraisals:
   i. Development of the business case and master planning.
   ii. Producing the facility cost model allowing option appraisals within the master planning phase and functionally based cost planning.
   iii. Executing the business case sensitivity analysis to test each option's rates of return on investment.
   iv. International benchmarking of airport indicators (operational and construction measures) to assist with target setting.
2. Advising clients on project brief, preferred procurement route and cash flow:
   i. Design of the incentivised procurement strategy and resultant contract terms and conditions involving the development of an innovative strategy and framework agreement against which contracts could be let.
   ii. TechT facilitated the target agreements, i.e. negotiated the target and any bespoke wording; the share per supplier within the Agreement was agreed by the suppliers themselves.
   iii. TechT and Laing O'Rourke provided comprehensive commercial benchmarking across the whole T5 programme to enable BAA to judge whether the anticipated final costs (AFCs) provided good/poor value for money compared to other BAA and non-BAA projects.
3. Analysing whole-life costs: implementation and management of an innovative value improvement process which secures cost and time based on themes of designing, buying and delivering better.
4. Planning the construction process:
   i. Project management of the planning supervision process.
   ii. Project planning undertaken jointly between BAA/Laing O'Rourke based on overall T5 Strategic Plan using Primavera; however QS role minimal.
   iii. Sub-project planning – each sub-project (£75–100 m) has its own cost and programme target – progress against the programme is commercially monitored.
5. Monitoring control of cost during pre-contract stage:
   i. TechT cost managers engaged with suppliers' cost managers to verify the cost plans and ensure alignment with the sched-

ule. The aim was to achieve a cost plan that was 95% bottom up, based on figures from suppliers by BAA's 'D Day' (the milestone before the site assembly starts but when most of the manufacture is complete and design is 95% complete).

   ii. TechT role is to use the data to help assess the cost to complete and advise on alternative decisions for alternative solutions if required.
   iii. TechT prepared benchmarking at facility, system and component level to demonstrate best value.
   iv. Contractors were required to monitor their pre-contract costs and continually update BAA on the forecast cost for this period.

6. Preparing tender and contractual documentation:
   i. TechT supported BAA's supply chain team to negotiate and periodically review the commercial model agreements (CMAs) between BAA and each of the first-tier suppliers.
   ii. TechT supported BAA's supply chain in providing advice on appropriate procurement routes and choice of supply for each work package then utilised the CMAs as a basis for agreeing the AFC.
7. Advising on payments to contractors, cost control and settlement of final accounts:
   i. Part of TechT's role in advising on payments was to assure that only properly incurred costs were paid. This was part of TechT's independent assurance role. Although it is an actual cost contract there is a need to provide assurance that only properly incurred costs are reimbursed. This requires commercial tension to be employed.
   ii. Majority of first-tier contractors reimbursed on 'an actual cost based' form of contract; interim payments and final account based on actual cost.
   iii. TechT work in conjunction with BAA's cost verification team to check, audit and then verify that the costs approved for payment are valid.
   iv. Laing O'Rourke used COINS (accounting package) to collate costs; BAA/TechT had *read-only* access to COINS to verify that costs were properly incurred.
   v. BAA ran the project using Oracle/Artemis; contractors feed information into Artemis using *comma separated variable* (CSV) files on a weekly basis, updating costs, progress and forecasts on a work breakdown structure (WBS) basis.
   vi. Laing O'Rourke developed their own database system of weekly capturing costs against WBSs, recording progress and forecasting final costs; this was loaded electronically into BAA's Artemis system every Wednesday for the previous week.

    vii. There are almost 1800 accounts to be finalised – TechT supported the BAA commercial team in developing the final account process and programme and the target is that 95% of all accounts will be agreed by the time T5 opens.
8. Controlling the project on behalf of their employer:
    i. Supporting BAA's integrated team in the preparation of the project process and procedures and implementation of a project-control system and software including change control and risk management.
    ii. Introduction of performance management system with KPIs based on cost, time and quality criteria.
    iii. Provision of monthly earned value analyses (linking time and cost) and schedule performance indices.
    iv. TechT have driven the production of and are the acknowledged owners of two of the ten core processes at T5 – cost management and commercial management.
    v. TechT collated the majority of other project information from their databases to enable benchmarking to be compiled at facility, system and component level in order to demonstrate value for money and to set appropriate targets.
9. Negotiating with client or subcontractors:
    i. TechT work with BAA in agreeing AFC targets throughout the duration of the programme; each project and sub-project has its own AFC target, progress against which is reported on and discussed on a monthly basis.
    ii. TechT together with BAA's Supply Chain team and first-tier contractors work as collaborative teams.
    iii. TechT supported BAA Supply Chain team in setting up 'buy clubs' in M&E services and fit-out to promote value procurement of suppliers and materials; TechT were the facilitators and main negotiators.
10. Reporting on the programme and financial matters:
    i. The TechT cost managers are integrated within their project and sub-project teams; they report to the project leader, who in turn reports to the T5 directors; TechT are also represented at programme office level.
    ii. The heads of cost management, commercial management and performance measurement are all senior TechT people reporting to BAA directors.
11. Risk and value management:
    i. Ongoing option appraisal and value engineering of construction systems.
    ii. Ongoing value improvement initiative focusing on productivity improvements and global acquisition combining leverage and partnering based buying.
    iii. Development of bespoke risk management process aligned to the T5 insurance cover.

12. Giving contractual advice in case of dispute: minimal dispute due to collaborative nature of the contract.

On the T5 project, TechT, as part of the integrated commercial team, have embraced many areas of best practice that can be brought to future projects, including:

- Best-in-class integrated information and knowledge-capture system that consolidated performance data from all suppliers into a single information repository.
- Developing and delivering leading project control practices and the creation of integrated teams. EC Harris and Turner & Townsend were working together under a performance framework to drive improvement. They adopted a common approach that aligned with BAA's own practices and benefited from a well staffed, highly motivated team.

The benefits flowing from these best practices included: a streamlined management team; improved resource management and development; sustained service delivery at ever improving quality levels over more than a decade; successfully embedded project control processes and systems throughout BAA Heathrow and across the supply chain; and a cheaper overall solution through cost containment of resource base and process improvement.

This expertise has enabled the TechT collaborative partners to join forces in an innovative initiative designed to bring significant benefits to clients in the nuclear sector. In August 2006 it was announced that a new joint venture company, Nuclear JV, comprising EC Harris LLP and Turner & Townsend Project Management Limited, had achieved its first success with the appointment by British Nuclear Group to act as the sole service provider of project controls for the Sellafield development.

The contract will run up to 4 years and involve the provision of over 100 project control professionals, working in an integrated team with the staff of British Nuclear Group with the aim of developing the quality of service whilst reducing the overall cost. The nuclear industry is facing significant challenges particularly in connection with clean up and decommissioning, not to mention new build; the Nuclear JV is thus a significant development for the future.

EC Harris and Turner & Townsend are also working with BP to develop and implement a project controls best-practice manual for use by their global project controls community. Additionally they are assisting the Shell Group to develop a standard global platform for Norms, Estimating, Contracts and Project Controls. This will enable Shell to benchmark each site and drive performance improvements on a global basis by leveraging knowledge capture and adoption of best practice (http://www.thenjv.com).

Conclusions and recommendations

The T5 project was the watershed in embracing the principles of *lean construction* in the UK and has required a complete change in the mindset and culture of the participants. The client had a huge role to play in the project success. Instead of writing into its contracts penalties for failure BAA accepted all the risk from the outset and guaranteed its suppliers an agreed margin, thus sending out a positive message to the whole project team.

BAA created a single entity harnessing the 'intellectual horsepower' working to get the job done rather than poring over contracts to find excuses. In return for its goodwill BAA demanded absolute transparency in the books of its suppliers; this created a collaborative approach in which all team members were equal and which encouraged problem solving and innovation in order to drive out unnecessary costs. It also removed the need for contractual claims and litigation, and drove up productivity levels.

Within the post-Egan era the role of the QS, or *cost consultant*, has blossomed becoming far more client focused. This new role requires an in-depth understanding of strategic project management embracing best practice within the whole project cycle. The profession now demands innovative problem solvers with high-level information technology skills that can contribute to the success of the project as part of an integrated team increasingly in an open-book target cost environment. This approach makes for a more challenging environment in that commercial assurance must be given within a more collaborative project environment.

BAA's enlightened approach created a collaborative environment, which led to the implementation of industry best practices and world-class performance. This approach was particularly relevant to long-term projects with high risk and high complexity, valued at £200 million and above, but might not be so relevant for smaller more straightforward projects.

This chapter has demonstrated that the quantity surveying profession has embraced the changes and challenges of the new era. TechT, working as part of an integrated team on T5, enabled BAA to introduce best practice into cost and commercial management within the broad philosophy of lean construction. Other major blue chip clients are increasingly demanding similar collaborative, innovative solutions for their global systems of project control.

Big issues such as world environmental, sustainability and carbon dioxide emissions are a real concern and will have an impact on everyone. Significantly, Davis Langdon & Seah report in their 2005–2006 *Global Review* that they are introducing sustainable design thinking and sustainable metrics into their cost plans, thus creating further opportunities for the profession (http://www.davislangdon.com).

Finally, international construction and management consultants acknowledge that their main asset is their people. Attracting and inspiring people will remain the major challenge for the construction professions. Perhaps the move away from adversarial to increasingly collaborative relationships, dictated and enforced by major clients like BAA, will help to provide better projects for clients and stakeholders, and a more challenging and exciting environment for the staff associated with such projects.

Acknowledgement

The author acknowledges the invaluable assistance offered by Mike Knox from EC Harris and Peter Ellis from Laing O'Rourke in compiling the original COBRA2006 paper together with information accessed on the Turner & Townsend website and final check for accuracy by Antonia Kimberley at BAA. Additional information for this chapter was kindly provided by Mike Cotten at Turner & Townsend.

References

Ashworth, A. and Hogg, K. (2000) *Added Value in Design and Construction*. Longman, Harlow, Essex.

BAA T5 factsheet *The key stages of Terminal 5*. (www.baa.com/t5 – accessed 5 July 1996)

Bayfield, R. and Roberts, P. (2004) Insights from beyond construction: collaboration – the Honda experience. *Paper presented to the Society of Construction Law*, 15th June (www.scl.org.uk – accessed 15 September 2006).

Boultwood, J. (2005) *Heathrow T5: a case study*. NEC Annual Seminar, 26 May, ICE London.

Broughton, T. (2004) Terminal 5 Supplement: a template for the future how Heathrow Terminal 5 has rebuilt the building industry. *Building*, 27 May.

Cartlidge, D. (2002) *New Aspects of Quantity Surveying Practice*. Butterworth Heinemann, Oxford.

Construction Industry Board (1996/97) Construction Industry Board Reports WG1–WG12, Thomas Telford.

Davies, R. (2006) The QS transformation. *RICS Business*, March.

Douglas, T. (2005) Interview: Terminal 5 approaches take-off. *Times Public Agenda Supplement*, 6 September.

Egan, Sir John (1998) *Rethinking Construction*. Department of Environment, Transport and the Regions, London.

Ferroussat, D. (2005) *Case Study BAA Terminal 5 Project – The T5 Agreement*. BAA controlled document, publisher unknown.

Geoghegan, M. (2005) *Quality Governance and Management of Heathrow T5 Construction*. Meeting London Branch IQA, May (http://www.london-sw-branch-iqa.org.uk/meetings.htm – accessed 30 September 2006).

Latham, Sir Michael (1994) *Constructing the Team*. HMSO, London.
Morris, P.W.G. and Hough, G.H. (1987) *The Anatomy of Major Projects: a study of the reality of project management*. Major Projects Association, John Wiley & Sons, Chichester.
National Audit Office (2001) *Modernising Construction*. Report by the Comptroller and Auditor General, HC 87 Session 2000–2001: 11 January, The Stationery Office, London.
National Audit Office (2005a) *Improving Public Services through Better Construction*. The Stationery Office, London.
National Audit Office (2005b) *Improving Public Services through Better Construction: Case Studies*. Report by the Comptroller and Auditor General | HC 364-II Session 2004–2005 | 15 March, The Stationery Office, London.
Office of Government Commerce (2003) *Achieving Excellence in Construction Procurement Guidance Pack* (www.ogc.gov.uk).
Office of Government Commerce (no date) *Managing Risks with Delivery Partners: a guide for those working together to deliver better public services*. Treasury, London.
Riley, M. (2005) Interview. *Turner & Townsend News Issue 31*, Turner & Townsend, London.
RICS (2002) *APC/ATC Requirements and Competencies, RICS Practice Qualifications, July 2002 – edition one*. RICS, London.
Simpkins, E. (2005) Terminal velocity. *RICS Business*, October, 16–20.
Strategic Forum for Construction (2002) *Accelerating Change*. The Construction Confederation, London.
Whitelaw, J. (2004) Taking control. *New Civil Engineer*, NCE Terminal 5 Supplement, February, *http://www.turnerandtownsend.com: case studies: Terminal 5 Heathrow Airport – role of the cost consultant* (accessed 6 May 2006).
www.wmcoe.gov.uk/download.asp?id=450 – accessed 30 September 2006.
http://www.thenjv.com/pdf/An%20Introduction%20to%20the%20Nuclear%20JV.pdf – accessed 27 January, 2007).
http://www.davislangdon.com/pdf/Global_review_06.pdf – accessed 27 January 2007.

3 Client requirements and project team knowledge in refurbishment projects

Cynthia ChinTian Lee and Charles Egbu

Introduction

Given the characteristics of refurbishment projects, the refurbishment process can be complicated, requiring a wide variety of project team skills. Clients are very reliant on the knowledge and skills of the project team in refurbishment projects. It is argued that a knowledgeable project team is able to understand and interpret client requirements. A knowledgeable team also has an increased chance of delivering a project on time and within budget.

Frequently, project team members only access the knowledge available when the process of executing a task comes to a halt in the face of a problem. There is a general lack of understanding of how valuable the fusion of processes and knowledge can be. Actually taking the distilled knowledge and making it available to people executing the process is somehow overlooked (Russell Records, 2005).

A perceived low level of client satisfaction has long been acknowledged as an important issue to be addressed in construction, for example by Bennett *et al.* (1988), Latham (1994), Egan (1998) and Smyth (2000). This chapter presents a methodology for the matching of knowledge of the project team with client project requirements during the refurbishment process in the construction industry.

The need to meet client project requirements

Successful projects are characterised by meeting client requirements. According to Mbachu (2003), client satisfaction adds value to the organisation from a number of perspectives: creation of sustainable client loyalty to the firm; repeat purchase; acceptance of other products/

services offered by the service firm; increased market share and profitability levels; creation of positive word-of-mouth; and a measure of market performance. Satisfaction is also important to the client because it reflects a positive outcome from the outlay of scarce resources and/or fulfilment of unmet needs (Day and Landon, 1977). Dissatisfaction, on the other hand, leads to undesirable consequences, such as negative word-of-mouth, complaints, redress-seeking, reduction of market share and profitability levels (Oliver, 1981), and possible divestment from the industry (Kotler, 1997).

One of the reasons cited for refurbished facilities falling short of client expectations is that construction professionals usually design with the needs of the environment, aesthetics and posterity in mind, not so much that of the client (Latham, 1994). A reason advanced by Mbachu (2003) for the prevalence of client dissatisfaction within the construction industry is that the area of client needs and satisfaction in the industry is still little researched. Not much effort has been made to identify the needs and expectations of clients, which is crucial to ensuring client satisfaction. It is, however, accepted that there are instances where clients' needs are met but clients are still dissatisfied because their expectations were higher than their needs alone might suggest. Their expectation had an element of desire also. Green and Lenard (1999) noted that the problem is a recurring one throughout the global construction industry and the industry has invested little time and attention in investigating the needs of its clients compared to other economic sectors. As project teams in the construction industry consist of participants from many disciplines, achieving client satisfaction is a collaborative effort of the project team. To ensure that a client's project requirements are fully met, the appointment of capable expert consultants is important and knowledge, especially shared knowledge within the project team that is applied through collaborative working, provides an important contribution to overcome the problem of client dissatisfaction in the construction industry.

Knowledge of the project team

Martensson (2000) regards knowledge as something that resides in people's minds and as one of the most important resources to an organisation (Nonaka and Takeuchi, 1995). For professional and technical service firms, the reputations, experience and skills of employees are their main assets and knowledge is an essential resource (Empson, 2001) and a primary source of competitive advantage (Grant, 1996; Galunic and Rodan, 1998). Knowledge also includes knowledge associated with an understanding of the client organisation (Smyth, 2000).

Very often, knowledge is confused with information. A distinction made between information and knowledge by Ash (1998) and Kirchner (1997) is that information has little value and will not become knowledge until it is processed by a human mind and that knowledge involves the processing, creation, or use of information in the mind of the individual. Although information is not knowledge, it is an important aspect of knowledge (Martensson, 2000).

Sustained competitive advantage is obtained through capabilities and resources that are valuable, rare, non-imitable and non-substitutable (Barney, 1991). Individual expertise and human capital have been identified as a key resource to organisations because they add value to the firm, they are unique and rare to competitors, imperfectly imitable, and cannot be substituted with another resource by competing firms (Wheelwright and Clark, 1992). Deploying experts into project teams is therefore the mechanism by which professional service organisations coordinate and apply the diverse expertise and experience embedded within individuals (Teece, 1998), creating a knowledge network within the team, although sharing knowledge between team members may also create some tensions concerning competitive positions of the team members. The ability to match expertise with client needs and expectations is key to improving the competitive advantage for project-based professional or technical service organisations (Maister, 1993) where collaborative knowledge gained through successive projects is captured by organisations to continually enhance their market position. From a knowledge-based strategic management perspective, the creation of an optimal mix of a project team (i.e. having expertise in its membership drawn from across organisations) that has requisite skills and competences matched to the client project requirements, will lead to client satisfaction with the project.

The importance of a knowledgeable project team has been indicated by Othman *et al.* (2005). Being the originators of brief development, project team members' knowledge or the lack of it can be a source of value or risk to the project. This view is echoed by Hatten and Lalani (1997) who suggest that by selecting an appropriate consultant team the chance of delivering a project on time and within budget might increase. Cooley (1994) concurred and felt that good consultants would bring genuine and lasting value to the organisation they serve.

Ideally, a multi-disciplinary design team is staffed in such a way that both the levels and the distribution of knowledge within the team match those required for the project (Fong, 2003). However, the knowledge or expertise of staff is seldom deployed according to the requirements of the project as a result of a shortfall in knowledge, such as the mismatch between staff expertise and project domain knowledge or because of *ad hoc* staffing approaches followed in most organisations due to a sudden increase in workload (Fong, 2003). It is also the case, that in order to manage transaction costs more efficiently, firms may also deploy staff in terms of their availability regardless of their knowledge.

The need for knowledge in refurbishment projects

In refurbishment work, there are many tasks where decisions are shaped not only by external factors (e.g. regulations, materials price rises) but also by the experience-based capabilities and future workload of the firm's personnel and its policy on the allocation of resources. In such situations, one is likely to find that experts rely on relatively unstructured methods in arriving at a decision.

Knowledge is the ideas, wisdom and facts managers acquire through experience, theory and practice; the acquisition of which gives them an ability to understand. Knowledge can be potential or manifested in performance.

In refurbishment, with the increase in contract labour, together with a corresponding increase in fragmented specialised work and the difficulties associated with labour on site, the skills of leadership and communication become even more necessary. Also, with an increasing need for speed of response to address the issues arising from variations to the works, the skill of communication becomes vitally important.

Given the complexities of refurbishment projects, refurbishment work demands greater supervision than new build work (Koehn and Tower, 1982). Willenbrock *et al.* (1987) are of the view that the nature of refurbishment work, coupled with a long working week and overtime work by construction personnel, leads to low morale and low productivity of refurbishment work. Thus, the skill/knowledge of motivating others is needed to boost the low morale and low productivity of refurbishment work.

Demolition work can involve the disposal of hazardous substances such as asbestos and lead. Statistics from the Health and Safety Executive (HSE, 1998) show that the repair and maintenance sector, including refurbishment, accounts for about 43% of the total number of construction fatal accidents in the UK. The need to understand and be able to control substances hazardous to health, such as asbestos and lead, especially by the site management team, is also of the utmost importance. Egbu (1996) argued that appropriate management strategies need to be developed to cope with the safety risks and hazards, especially for works carried out with tenants in occupation.

Refurbishment work is characterised by high risk, uncertainty and high numbers of variation orders to the works. Working under such situations, and at the same time attempting to achieve the stipulated time for project completion, managers would have to make impromptu and sound decisions. The skill of decision making, therefore, is of great importance at all levels of management, which is enhanced by information and knowledge from individual project organisations and knowledge that has been generated and shared by the team from the outset. In this environment of uncertainty, where costs are likely to escalate and controlling the financial parameters of the refurbishment process is necessary, the skills and knowledge associated with financial forecasting and planning are particularly important.

With refurbishment works often involving working in confined sites, the knowledge of site organisation is important. In addition, there is a need to understand the nature and qualities of the materials used originally so as to match them exactly or look for a material which blends in and harmonises with the existing environment.

The high level of uncertainty in refurbishment work tends to lead to project time over-run. The skills and knowledge associated with managing time therefore are also necessary. The relatively high degree of importance attached to managing time is supported by Jothiraj and Fellows (1986), who observed that time performance was the major factor in determining overall client satisfaction with commercial refurbishment projects. However, every refurbishment project is different and understanding the client includes the weighting given to time, cost and quality criteria.

The ability to cope with unexpected changes, conflicts and crisis is needed in refurbishment work. The skill/knowledge associated with the analysis of project risk/uncertainty is also of high importance. The high degree of importance attached to project risks/uncertainty by all levels of management reflects the high levels of risk and uncertainty associated with refurbishment works (Quah, 1988; Teo, 1990). Refurbishment work therefore demands requisite skill/knowledge associated with being able to assess and analyse risks/uncertainty in construction work. Project team members with diverse skills, knowledge and experience are required to work together to resolve issues or problems encountered in a project; with contributions from the knowledge network across the team membership and, ideally, through collaborative knowledge working as a project team, client's satisfaction can be better achieved.

Conceptual framework

According to Eisenhart (1991), a conceptual framework is an argument including different points of view and culminating in a series of reasons for adopting some points, that is, some ideas or concepts and not others. The adopted ideas or concepts then serve as guides: to collecting data in a particular study, and/or ways in which the data from a particular study will be analysed and explained.

In this section, key issues identified in the literature review are summarised and synthesised in order to identify theoretical gaps, leading to the identification of the research issues to be explored in this chapter. The conceptual framework for matching project team's knowledge with client project requirements, as illustrated in Figure 3.1, shows a notional end-to-end 'product' development process and ensures that there are

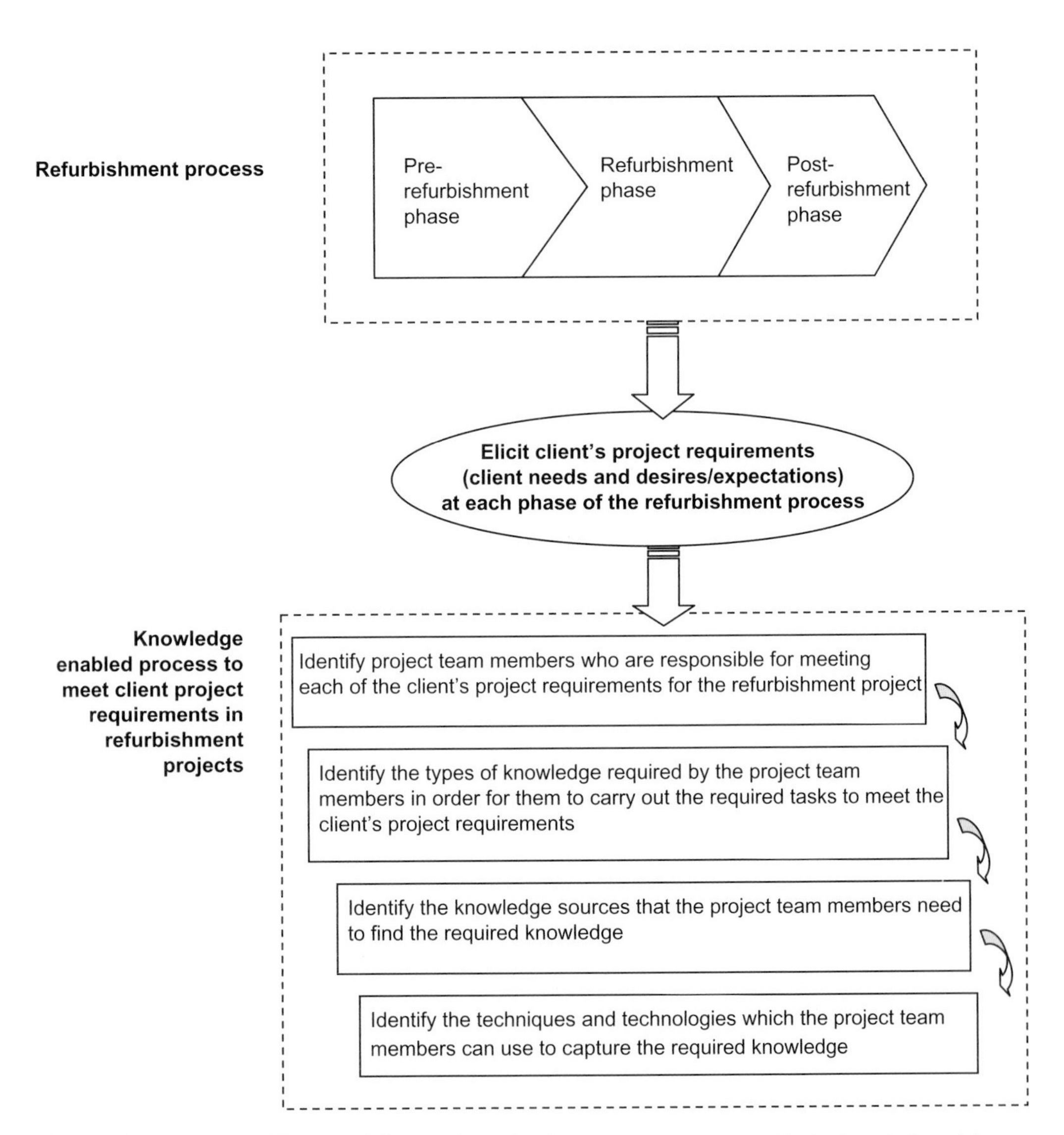

Figure 3.1 Conceptual framework for matching client's project requirements with the knowledge of the project team for refurbishment projects

means in place for organising what data, information and knowledge are available for each task in the overall refurbishment process.

One of the major causes of process delays is the lack of the right information or knowledge at the right time, which causes the process execution to slow down or come to a halt. Until appropriate knowledge is acquired for the performance of tasks in the process, there will be a delay in the process and incurrence of costs. Thus, there is a need to integrate knowledge management into the refurbishment process and for this the creation of a knowledge map is an essential task.

According to Grey (1999), the types of questions to be asked to develop a knowledge map include:

- What type of knowledge is needed to do your work?
- Who provides it, where do you get it, how does it arrive?
- What do you do, how do you add value, what are the critical issues?
- What happens when you are finished?
- How can the knowledge flow be improved, what is preventing you doing more, better, faster?
- What would make your work easier?
- Who do you go to when there is a problem?

As Figure 3.1 illustrates, client project requirements are elicited at each phase of the refurbishment process. In order to satisfy client project requirements, the first process involves identifying the project team member who is responsible for meeting each of the client project requirements. The second process is the identification of the types of knowledge that are required for the project team member to carry out the tasks in the refurbishment process. Recognising what knowledge is required will allow an accurate search for the required knowledge. The third process involves the search for knowledge and the sources of knowledge to which the project team member can go in order to acquire the requisite knowledge. Having identified the types of knowledge and the sources from which knowledge can be acquired, the fourth and last process looks into the methods employed to capture the required knowledge.

The research project upon which this chapter is based set out to understand:

- Client project requirements for refurbishment projects.
- The types of knowledge required to meet client project requirements.
- The sources of knowledge used by project team members to meet client project requirements.
- Knowledge capture techniques and technologies used by project team members to meet client project requirements for refurbishment projects.

Research methodology

Methodology is concerned with the logic of scientific inquiry; in particular with investigating the potentialities and limitations of particular techniques or procedures (Grix, 2002). Very often methodology is confused with methods; the latter is the techniques or procedures used to collate and analyse data.

The research associated with this chapter is qualitative for the following reasons:

1. Refurbishment projects are complex and rely heavily on communication and interaction.
2. The needs and requirements of clients and users vary due to differences in perceptions of different clients/users.
3. The research involves the investigation of uncertainty and dynamic change in refurbishment projects.
4. The knowledge of people is intangible.
5. The aim of the research is to introduce change into the process of the development of the brief.
6. In order to elicit the knowledge creation process of project team members, participation from stakeholders is required.

The research paradigm used in this context is phenomenological: understanding meanings, or looking at, describing and understanding experience, ideas, beliefs and values in the construction industry. In order to gain insights to the research area, make inferences and draw conclusions from the research, methodological triangulation was adopted. To optimise the quality of data collection, non-probability sampling was the chosen primary sampling method.

Survey research methodology was adopted in this research to investigate a particular phenomenon – the knowledge-enabled process for meeting client project requirements for refurbishment projects in the construction industry.

Research methods

The literature review carried out has generated the following research questions:

1. What method(s) is employed by project teams or clients in matching client project requirements with the requisite knowledge of the project team?
2. Do the types and complexities of refurbishment projects have an impact on the types and sources of knowledge that the project team

draws upon to address specific refurbishment tasks in order to meet client project requirements?
3. Do the types and complexities of refurbishment projects affect the choice of knowledge capture tools and techniques employed by the project team to meet client project requirements?

An examination of the research questions justified the use of questionnaire survey and semi-structured interviews as the method for data collection.

Sampling

The study population in the data-gathering stage consisted of:

1. The client population: developers, hotel owners and private and public hospitals.
2. The project stakeholders: professionals who are involved in refurbishment projects. The key professionals include architects, quantity surveyors, structural engineers, building services engineers and contractors.

The sampling frame for the questionnaire and interview included client, users, architects, quantity surveyors, consulting engineers and project managers who have the experience or are currently working on hotel, retail, offices and hospitals refurbishment projects. The choice of this sampling frame was based on the generic organisational structure of the UK construction industry (Riley and Brown, 2001) which illustrates the range of contributors that are required for a construction project.

The sampling units (respondents) were developers, facilities managers of hotels and facilities managers of private and public hospitals. For project stakeholders, the respondents were drawn from registered architects from the RIBA list, quantity surveyors registered with the RICS, structural engineers and building services engineers from the Institute of Structural Engineers and Chartered Institution of Building Services Engineers respectively. The contractors were selected from a list of refurbishment contractors searched through online databases.

Out of the 600 questionnaire surveys sent out between February and March 2006, 124 completed questionnaires were returned, representing a response rate of 20%; this falls within the norm of 20–30% response rate for most postal questionnaire surveys conducted in the construction industry. Forty interviews were conducted between May and July 2006.

Findings and analysis

Matching client project requirements with knowledge of the project team

The literature review conducted indicates that a firm's technical capability, experience on similar projects and financial soundness are the main selection criteria which clients place importance on when selecting their project team (consultants and contractor). To meet client's selection criteria, organisations have to demonstrate that they have employees with the required technical capabilities and past experiences for the project during the pre-selection stage. In order to find out how organisations identify individuals with the relevant technical capabilities and past experiences within the organisations, respondents in this research study were asked the approaches that are adopted by their organisations to identify individuals that have the requisite knowledge to meet client project requirements.

Table 3.1 documents the views of the respondents on the approaches adopted as well as the extent of use of these approaches. The approaches are grouped under two main categories: formal and systematic. Formal is defined as procedures that are carried out in accordance with established forms, conventions or requirements. Systematic, on the other hand, are procedures carried out in a step-by-step manner.

The analysis of interview data conducted with 40 industrialists who have experience working in refurbishment projects shows that the main approach adopted by companies to match client project requirements

Table 3.1 Approaches adopted by organisations to match client project requirements with the knowledge of the project team

Approaches adopted by organisations	Extent of use
Formal approaches	
Look at past experiences of an individual	22%
Curiculum vitae of an individual	17%
Company's intranet	14%
No set up procedures	11%
Recommendations and references from others	6%
Decision made by top management	3%
Interview with the individual	3%
Conduct staff appraisal	2%
Have a network of consultants that are used regularly	2%
Systematic approaches	
Use software like: mindmap, spreadsheet, expert locators, ACID Test	12%
Conduct performance review at the end of the project	2%
Use pre-qualification questionnaire	2%

with the knowledge of their existing staff is by looking into the individual past experiences and curriculum vitae/résumés. This information is usually available and stored in the company's intranet or database.

From Table 3.1, it can be seen that the approaches adopted by organisations are mainly formal procedures. A large number of organisations (11%) do not set up any procedures at all and 3% of the respondents indicate that the allocation of staff to a project is decided by the top management. When it comes to selecting their consultant team, many clients depend on recommendations from others who have favourable previous experience with the consultant. A small number of respondents (2%) have a network of consultants that they use regularly for their projects because these consultants are familiar with their work procedures and therefore are expected to better meet their needs.

Types of knowledge

In this research study, respondents were given a list of client project requirements for refurbishment projects in the questionnaire and asked to indicate the extent to which the types of knowledge are used to address each of the project requirements. A choice of two types of knowledge is presented – tacit or explicit knowledge. Tacit knowledge is highly personal and hard to communicate; it is deeply rooted in individual experience whilst explicit knowledge can be easily communicated and shared and can be acquired through factual statements such as material properties and technical information (Koskinen *et al.*, 2003).

A comparison between the extent of use of tacit and explicit knowledge by project team members in their attempt to meet client project requirements indicates that the use of tacit knowledge is 13% higher than explicit knowledge. The Kruskal Wallis tests conducted on the questionnaire survey data to look at the relationship between years of experience and the types of knowledge revealed that project team members with more years of experience tend to use more tacit knowledge than explicit knowledge. This verifies that tacit knowledge is accumulated and acquired through 'learning-by-doing'.

Referring to the Kruskal Wallis test results shown in Table 3.2, the extent of use of tacit knowledge is not significantly different between the different types of refurbishment projects. However, there is a significant difference across the different types of refurbishment projects and the extent of use of explicit knowledge. An interesting phenomenon to note is the role played by the project team and their use of types of knowledge. The extent of use of tacit knowledge is not affected by the role played by a project team member; however, explicit knowledge varies according to the role played by a project team member. Since explicit knowledge is knowledge that can be found from documents and electronic databases, this result therefore indicates that different project

Table 3.2 Kruskal Wallis test result on the relationship between type of refurbishment projects and tacit/explicit knowledge

	Chi-square	Degrees of freedom (df)	Level of significance	Mean rank				
				Retail	Offices	Hotels	Hospitals	Others
Tacit knowledge	0.483	4	0.975*	64.85	58.85	55.71	58.44	57.46
Explicit knowledge	18.739	4	0.001*	54.25	67.49	44.00	78.47	44.45

* Results are statistically significant at $p < 0.05$

team members use different sources of knowledge to meet client project requirements.

Respondents were asked if the complexities of refurbishment projects affect the types of knowledge used to manage refurbishment projects. About 43% of the respondents from the semi-structured interviews and 62% from the questionnaire survey indicate that the complexities of refurbishment projects do affect the types of knowledge used to manage refurbishment projects. Additionally, respondents also state that the types of knowledge used to manage refurbishment projects are dependent on the requirements of the refurbishment project. Put differently, the use of knowledge is context specific to the refurbishment project.

Sources of knowledge

Sources from which knowledge is found can be broadly categorised under four modes: oral, paper, electronic and others (includes: personal experience, seminars and exhibitions). The research results presented in Table 3.3 show that the interaction with colleagues and personal experiences are the main sources of knowledge that respondents draw upon to meet client's requirements in refurbishment projects. It is also revealed from the questionnaire survey and interview that knowledge from paper-based sources is popular in the refurbishment sector. However, publications, trade magazines and catalogues are relatively more important in keeping up to date whereas conversations with clients, vendors and external colleagues, and company in-house documents are relatively more important in problem solving.

The use of electronic information systems is changing rapidly, as more project managers and engineers obtain access to internal networks and internet. From the research results, the electronic mode of obtaining knowledge is the second least used by the refurbishment sector as a source of knowledge. This could be explained by the nature of the con-

Table 3.3 The extent of use of various sources of knowledge

Mode	Sources of knowledge	Response from questionnaire survey (%)	Response from semi-structured interview (%)
Oral	Interaction with colleagues	50	27
Paper	Publications/magazines/catalogues (internal/external)	7	22
	Company/in-house documents	9	
Electronic	Internet/intranet	3	20
Other	Personal experiences	30	31
	Seminars and exhibitions	1	

struction industry where the adoption of information technology has been slow and being populated by mostly small–medium enterprises which have a weakness in technological competencies; using electronic means of obtaining knowledge can be restricted.

From the questionnaire survey results, it is observed that seminars and exhibitions are the least popular means from which knowledge is sourced. One explanation for this is that seminars and exhibitions are often considered as continuing professional development (CPD) training and not as a source for acquiring knowledge for immediate use. In addition, organisational support for funding and time for CPD training is an issue yet to be fully recognised by some organisations as an activity worthy of management time.

Techniques and technologies for knowledge capture

The findings revealed that 'face-to-face discussions', 'pre/post project review' and 'project documentation' are the knowledge-capture techniques most employed by respondents for acquiring the knowledge necessary to perform tasks to meet the list of client project requirements in refurbishment projects. Knowledge-capture technologies that are deemed to be used more often by the respondents are emails and project database. Respondents also indicate that project documentation is a more effective technique of knowledge capture because it is believed that this method would enable the dissemination of information which can be used by individuals and in turn internalised as knowledge. Colleagues within the organisations can have access to the recorded knowledge for future projects. Interaction with colleagues was seen as a knowledge-capture technique that is used personally and not for the benefit of the organisation. Similarly, emails are a more personal form of knowledge-capture technology whilst database and internet/intranet are technologies that organisations could make use of for knowledge sharing and dissemination.

Matching client project requirements with knowledge of the project team in refurbishment process

Although there is some literature written on the desirability of the customisation of teams to meet the requirements of projects and clients' needs (Maister, 1993; Teece, 1998; Smyth, 2000), there is little empirical work that has considered the methods of achieving this; and certainly not for refurbishment projects in the construction industry. There is no available method to match the knowledge of the project team with client project requirements during the refurbishment process.

The proposed methodology has been developed from the literature review, questionnaire survey and semi-structured interview results. It focuses on the matching of client project requirements in refurbishment projects and its purpose is to enable organisations to meet client project requirements by allocating individuals with the requisite knowledge to form a project team.

Figure 3.2 shows the developed methodology for matching client project requirements with the knowledge of the project team for refurbishment projects. The developed methodology has the important stages detailed below.

Identification and categorisation of client's project requirements

Client expectations and initial or general requirements are first elicited through the business development process, which is enhanced using relationship marketing techniques, although the knowledge is seldom transferred to inform the second stage (Smyth, 2000). Second, client project requirements are elicited using common methods like questionnaire, interviews, brainstorming exercise or discussion sessions conducted among the project team, end-users and client or client representative(s).

From the list of project requirements elicited, the requirements are categorised under the types of knowledge that are needed. The requirements are categorised as follows:

1. Building (knowledge of fabric of the building, the extent of deterioration, history of the building etc.).
2. Client (knowledge of what client's project requirements are).
3. Authorities (knowledge of authorities' requirements and regulations).
4. Management (knowledge on the management of refurbishment project).
5. Personal (knowledge acquired from own experience).
6. Industry (knowledge of what is available in the industry in terms of the technologies and materials).

Identification and categorisation of client's project requirements

(1) Elicitation of client's project requirements using techniques like interviews, questionnaire, brain-storming exercise and discussion sessions

(2) Categorise client's project requirements under the required knowledge types, for example:
- Client (knowledge of what client's project requirements are)
- Building (knowledge of fabric of the building, the extent of deterioration, history of the building etc.)
- Authorities (knowledge of authorities' requirements and regulations)
- Management (knowledge on the management of refurbishment project)
- Personal (knowledge acquired from own experience)
- Industry (knowledge on the available technologies and materials in the industry)

Establishment of knowledge gap and identification of knowledge sources

(3) Determine the knowledge required to meet client's project requirements. Assess the knowledge gap between the existing knowledge of the project team and the required knowledge by:
- Securitising each client's project requirements
- Identify what knowledge is required in order to meet the client's project requirements
- Formulate questions to ask if one has that required knowledge and conduct a questionnaire exercise

(4) Identify the knowledge source — **Knowledge sources** can be from: client, colleagues, project documents, publications, internet/intranet, past experience of colleagues or self

Knowledge capture

(5)

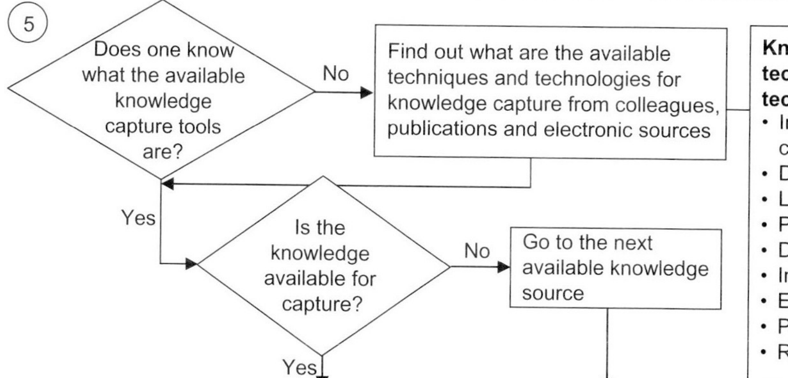

Knowledge capture techniques and technologies:
- Interaction with colleagues
- Documentation
- Lessons learned
- Personal record
- Database
- Internet/intranet
- Emails
- Photographic record
- Reviews

Capture knowledge from the identified knowledge sources using the selected techniques/technologies given the organisation's resources and usual practices

Matching of client's project requirements with project team's knowledge

(6) Fill knowledge gap between the required knowledge and the existing knowledge that the project team has with the captured knowledge

(7) Client's project requirements are matched with the knowledge of the project team

Figure 3.2 A process for matching client's project requirements with project team's knowledge for refurbishment projects (adapted from Kucza, 2001)

Establishment of knowledge gap and identification of knowledge sources

As refurbishment projects often involve a number of stakeholders, it is almost impossible to pinpoint a project team member that will meet specific client project requirements; instead the knowledge from various project team members may be required to meet just one of the client project requirements. Thus, the existing knowledge of the project team is assessed through a set of questionnaire exercises where the project team members are asked a series of questions relating to the client project requirements. This process will identify the knowledge gap between the required knowledge and the existing knowledge.

The sources from which knowledge can be drawn are also identified at this stage.

Knowledge capture

At this stage, knowledge is captured from the identified knowledge sources. From the research survey results, the knowledge-capture techniques and technologies used by the respondents are more for personal use. The choice of knowledge-capture techniques and technologies used tends to be personal or constrained by construction organisations' usual practice. The knowledge-capture techniques and technologies that are more commonly used by construction organisations for knowledge capture are project documentation, pre/post-project reviews, company's database and internet/intranet.

Matching client project requirements with project team knowledge

The knowledge gap between the existing knowledge of the project team and the required knowledge which was earlier discussed will be filled by the captured knowledge. Client project requirements are matched with the knowledge of the project team at this final stage.

Conclusion

Client satisfaction has long been acknowledged as an issue which needs to be addressed and the success of a project is often determined by the satisfaction of the client. In order to achieve client satisfaction two objectives have to be met: first, the translation of client needs into a design which specifies technical characteristics, functional performance criteria and quality standards and, secondly, the completion of the project within a specific time and in the most cost-effective manner. In order to meet the objective of client satisfaction, a knowledgeable project team that is able to understand and interpret client requirements and rely on their knowledge and experience to meet client requirements is likely to increase the chance of delivering a project on time and within budget.

In construction projects, each project team member brings with them a great deal of experience. Knowledge is also elicited from various project team members during the construction process. There are many people involved during the refurbishment process and each person is a potential source of knowledge. For refurbishment projects, where physical limitations of the building will always pose a problem when implementing client project requirements, project team members who have the knowledge and experience will be able to foresee problems and avoid possible variations and unnecessary work.

The findings in the research study unveiled the methods employed for matching client project requirements with the knowledge of the project team in refurbishment projects. The type of knowledge used by the project team is mainly tacit knowledge and the extent of use of such knowledge is influenced by the years of experience a project team member has in the industry and the complexities, typically, of refurbishment projects. The role played by a project team member does not affect the extent to which tacit knowledge is used. However, the extent to which explicit knowledge is used and the sources of explicit knowledge are influenced by the role played by a project team member. The research results show that interaction with colleagues and personal experiences are the main sources of knowledge that the project team members draw upon for problem solving in refurbishment projects. Other sources, like trade magazines, seminars and exhibitions, are knowledge sources used more for keeping up to date. The research analysis also indicates that the techniques and technologies used for knowledge capture by project team members can be differentiated between personal use and dissemination and sharing of knowledge.

The discussions in this chapter present a methodology to match the project team's knowledge with client project requirements during the refurbishment process in the construction industry. It describes a knowledge-enabled process for meeting client project requirements in refurbishment projects with the knowledge of the project team. In the proposed methodology, the processes adopted by the multi-disciplinary project team are not linear. Instead, they are interwoven processes where new or emergent knowledge is created or existing knowledge is combined to give new insights. This methodology sees the fusion of the refurbishment process and knowledge management and it is envisaged that it can provide a much higher return to construction organisations in terms of client satisfaction and effective work process.

References

Ash, J. (1998) Managing Knowledge Gives Power. *Communication World*, 15, 23–26.

Barney, J. (1991) Firm resources and sustained competitive advantage. *Journal of Management*, 17(1), 99–20.

Bennett, J., Flanagan, R., Lansley, P., Gray, C. and Atkin, B. (1988) *Building Britain 2001, Centre for Strategic Studies in Construction*. University of Reading, Reading.

Cooley, M.S. (1994) Selecting the right consultants. *HR Magazine*, 39(8), 100–103.

Day, R.L. and Landon, E.L. (1977) Towards a theory of consumer complaining behaviour. In: Woodside, A.G. Sheth, J.N. and Bennett, P.D. (eds.) *Consumer and Industrial Buying Behaviour*. North-Holland, New York, pp. 425–437.

Egan, J. (1998) *Rethinking Construction: Report of the Construction Task Force on the scope for improving the quality and efficiency of UK construction*. Department of the Environment, Transport and the Regions (DETR), London.

Egbu, C.O. (1996) *Characteristics and Difficulties Associated with Refurbishment*. Construction Papers No. 66, CIOB, Ascot.

Eisenhart, M. (1991) Conceptual frameworks for research circa 1991: ideas from a cultural anthropologist; implications for mathematics education researchers. In: Underhill, R. (ed.) *Proceedings of the Thirteenth Annual Meeting of Psychology of Mathematics Education – North America*. Psychology of Mathematics Education, Blacksburg, VA.

Empson, L. (2001) Introduction: Knowledge Management in Professional Service Firms. *Human Relations*, 54(7), 811–817.

Fong, P.S.W. (2003) Knowledge creation in multidisciplinary project teams: an empirical study of the processes and their dynamic relationships. *International Journal of Project Management*, 21, 479–486.

Galunic, D.C. and Rodan, S. (1998) Resource recombination in the firm: knowledge structures and the potential for Schumpeterian innovation. *Strategic Management Journal*, 19(12), 1193–1201.

Grant, R.M. (1996) Toward a knowledge-based theory of the firm. *Strategic Management Journal*, 17 (Winter Special Issue), 109–122.

Green, S.D. and Lenard, D. (1999) Organising the project procurement process. In: Rowlinson, S.M. and McDermott, P. (eds.) *Procurement Systems: A Guide to Best Practice in Construction*. E&FN Spon, London.

Grey, D. (1999) Knowledge mapping: a practical overview. http://www.smith-weaversmith.com/knowledg2.htm, accessed on 2 December.

Grix, J. (2002) Introducing students to the generic terminology of social research. *Political Studies Association*, 22(3), 175–186.

Hatten, D.E. and Lalani, N. (1997) Selecting the right consultant team. *Institute of Transportation Engineering Journal*, 67(9), 40–46.

HSE (1988) *Blackspot Construction: a study of five years fatal accidents in the building and civil engineering industries*. Health and Safety Executive, HMSO, London.

Jothiraj, T. and Fellows, R. (1986) Client control of commercial refurbishment projects. *Proceedings of the CIB 10th Triennial Congress*, 22–26 September, Washington, Vol. 7, 2837–2845.

Kirchner S.R. (1997) Focus on: database integration and management for call centres. *Telemarketing*, 16(2), 22–24.

Koehn, E. and Tower, S.E. (1982) Current aspects of construction rehabilitation. *ASCE, Journal of the Construction Division*, 108(C02), 330–340.

Koskinen, K.U., Pihlanto, P. and Vanharanta, H. (2003) Tacit knowledge acquisition and sharing in a project work context. *International Journal of Project Management*, 21(4), 281–290.

Kotler, P. (1997) *Marketing Management: Analysis, Planning, Implementation and Controls*, 9th Edition. Prentice Hall, New Jersey.

Kucza, T. (2001) *Knowledge Management Process Model*. VTT Publications, Espoo.

Latham, M. (1994) *Constructing the Team: joint review of procurement and contractual arrangements in the United Kingdom construction industry*. HMSO, London.

Maister, D.H. (1993) *Managing the Professional Service Firm*. The Free Press, New York.

Martensson, M. (2000) A critical review of knowledge management as a management tool. *Journal of Knowledge Management*, 4(3), 204–216.

Mbachu, J.I.C. (2003) *A critical study of client needs and satisfaction in the South African building industry*. Unpublished PhD Thesis, Faculty of Economic and Building Sciences, University of Port Elizabeth, Port Elizabeth.

Nonaka, I. and Takeuchi, H. (1995) *The Knowledge Creating Company: How Japanese Companies Create the Dynamics of Innovation*. Oxford University Press. New York.

Oliver, R.L. (1981) Measurement and evaluation of satisfaction process in retail settings. *Journal of Retailing*, 57(Fall), 25–48.

Othman, A.E., Hassan, T.M. and Pasquire, C.L. (2005) Brief development originators, value and risk sources to the project from the client's perspective. *5th International Postgraduate Research Conference*, 11–15 April, University of Salford.

Quah, L.K. (1988) *An evaluation of the risks in estimating and tendering for refurbishment work*. PhD Thesis, Heriot-Watt University, Edinburgh.

Riley, M.J. and Brown, D.C. (2001) Case study of the application of BPR in an SME contractor. *Knowledge and Process Management*, 8(1), 17–28.

Russell Records, L. (2005) The fusion of process and knowledge management. *Business Process Trends*, September 2005, http://www.bptrends.com, assessed on 2 December 2006.

Smith, H. (1975) *Strategies of Social Research: the Methodological Imagination*. Prentice-Hall, Englewood Cliffs.

Smyth, H.J. (2000) *Marketing and Selling Construction Services*. Blackwell, Oxford.

Teece, D.J. (1998) Capturing value from knowledge assets: the new economy, markets for know-how and intangible assets. *California Management Review*, 40(3), 55–79.

Teo, D.H.P. (1990) *Decision support and risk management system for competitive bidding in refurbishment contracts*. Ph.D. thesis, Heriot Watt University, Edinburgh.

Wheelwright, S.C. and Clark, K.B. (1992) *Revolutionising Product Development: Quantum Leaps in Speed, Efficiency and Quality*. The Free Press, New York.

Willenbrock, J.H., Randolph-Thomas, H. and Francis, P.J. (1987) Factors affecting outage construction efficiency. *Journal of Construction Engineering and Management ASCE*, 113(1), 99–116.

4 Contractual frameworks and cooperative relationships

Mohan Kumaraswamy, Aaron Anvuur and Gangadhar Mahesh

Introduction

Construction has been identified as a complex systems industry, where organising-by-projects, temporary coalitions of firms and a heavy client involvement in the product lifecycle are the norm (Shirazi *et al.*, 1996; Dubois and Gadde, 2002). Coping strategies allow the construction industry to deal with uncertainty and complexity (Dubois and Gadde, 2002; Shirazi *et al.*, 1996). Contained in most standard forms of contract, these coping strategies (see Chapter 5) create a pattern of 'tight' and 'loose' couplings and tend to result in adversarial relationships (Thompson *et al.*, 1998). The command-and-control nature of these strategies leads to a focus on short-term productivity at the expense of quality, innovation and learning. Performance incentives have been proposed as a remedy but their impact has been largely negative. Ashley and Workman (1986) found that although overall performance was marginally greater with their use, incentives created significantly more disputes and contractual disagreements. Emphasis is more recently being placed on alternative forms of control that facilitate trust-based relationships, although some of these later initiatives have been questioned (e.g. Cornes, 1996). No doubt, these developments are a testament to the commitment of the industry to improve the level and predictability of its performance.

However, there is concern over an emerging trend where cooperative relationships are emphasised more or less as the *end* instead of a *means* to achieve organisational goals. Cox and Thompson (1997) argue that the choice of a governance framework must be an objective, rather than a normative, decision. But determining what is a 'fit for purpose' contractual framework is problematic. There is no unified set of guiding theories on the procurement and management of projects. Fortunately, there is now a stimulating debate on the conceptual underpinnings of project procurement and management (e.g. Maylor, 2006). This debate will benefit greatly from a fundamental understanding of the decision-making processes of construction clients. The preoccupation with performance outcomes, to the neglect of *process*, clouds understanding and undermines the utility of any conceptual frameworks proposed to guide the procurement process. Such a fundamental understanding of process helps to refine and/or extend existing theories, or even to generate new theories (Eisenhardt, 1989; Gummesson, 2000).

This is the aim of this chapter: to contribute to a better understanding of clients' choices of contractual frameworks and working relationships, and the consequences of such choices. The rest of the chapter is structured in two main sections. The first section reviews existing or proposed conceptual frameworks, and proposes a suitable framework for construction contracting. The second section describes and draws on relevant findings from a case study to explain the rationale for, nature and implications of, the choices made by the client and the consequences thereof. The chapter ends with some concluding remarks.

Contractual frameworks

Relational contracting under Williamson's TCE framework

Agreement on a theory of contracting and construction project management is not easy (Koskela, 2003; Koskela and Ballard, 2006; Winch, 2006). However, it appears that the most dominant framework that is applied to construction is the transaction cost economics (TCE) framework. This is based on Williamson's (1979) seminal work, which was subsequently amended (Williamson, 1985), and other researchers building on his work. Williamson's work positions the firm as a nexus of contracts (Aoki *et al.* 1990). The basic argument of this stream of thought is that one of three organisational forms is most efficient for the governance of a firm's transactions. These are markets, hierarchies and relational contracting, which map on to price, authority and trust (Bradach and Eccles, 1989). Three dimensions of transactions – uncertainty, asset specificity and transaction frequency – constitute the evaluation criteria. These dimensions affect the costs associated with writing, executing and enforcing contracts. Hierarchies are more efficient than arm's-length market transactions in situations where high uncertainty, high asset specificity and frequent recontracting exist. The uncertainties about future performance and contingencies complicate the writing of contracts, the high asset specificity may lead to opportunistic bargaining and frequent recontracting is costly. In these situations a hierarchy (i.e. an authority relation, such as vertical integration) is most efficient (Williamson, 1985). Markets dominate at the opposite extreme. Intermediate levels of uncertainty and asset specificity lead to intermediate forms of control such as quasi-vertical integration. Williamson referred to such governance structures as 'relational contracting'. In relational contracting, the costs consist of the investments necessary to build trust through adherence to a norm of reciprocity (Bradach and Eccles, 1989). Obviously, these costs must be less than those involved in writing, executing and enforcing contracts.

The TCE approach has influenced research on and practice in business organisations. For example, the International Marketing and Purchasing (IMP) Group's 'interaction model' is based on Williamson's TCE framework (see Campbell, 1985). The TCE framework has also been elaborated in the context of construction organisations and applied in its original, amended and/or extended form to explain and analyse various aspects of the construction process (e.g. Winch, 1989, 2001; Cox and Thompson, 1997; Pietroforte, 1997; Thompson *et al.*, 1998; Walker and Chau, 1999; Lai, 2000; Costantino *et al.*, 2001; Turner and Keegan, 2001). The TCE approach has also been used in conjunction with other theories, notably the resource-based view (e.g. Bridge and Tisdell, 2004, 2006; Cox, 1996; Chang, 2006). For example, Cox (1996) amended Williamson's notion of asset specificity as related to sunk costs and market uncertainty to one related to the core competences of the firm. Cox argues that firms must defend their core competences at all costs if they are to survive and prosper. The make or buy decisions of firms should therefore be based on a 'relational competence analysis' (1996:61) – the functionality of the skills, expertise involved in, and the nature of, transactions, in ensuring the firm's sustainability (i.e. profitability) in the supply and value chain. The TCE approach recognises that idiosyncratic investment may lead to a (trust-based) relationship but provides scant information on how such a relationship develops. In other words, the stipulations of the TCE approach with respect to relational contracting, apply to situations where trust exists, but provide little information on how relationships grow and trust develops (this argument is revisited below). Indeed, Williamson's attempt to subsume the role of trust under relational contracting in the TCE framework has been criticised by many researchers as a stopgap approach (e.g. Oberschall and Leifer, 1986; Bradach and Eccles, 1989). In this regard, the TCE approach has been considered as being *reactive* (Pryke and Smyth, 2006; Smyth, 2006). Relationship marketing (see Gummesson, 1997, 2002), which is seen as a more *proactive* approach, is concerned specifically with creating high asset specificities of the kind that will, under the TCE framework, increase switching costs for buyers and, hence, close markets to competitors through relational contracting. In short, this is a strategy for competitive positioning adopted by suppliers to secure their profitability.

The concept of power

It is worth noting that not all TCE researchers in construction are in support of the idea of trust as underpinning the notion of relational contracting under the TCE framework. After noting that there is a lack of trust in the construction industry, Cox and Thompson (1997) questioned the relevance of trust in contractual relations. They suggest that power – ownership of and/or control over critical supply chain resources – and *ex ante* incentives, for example the promise of future work, are more viable options for facilitating cooperation. Cooperation in this chapter refers to the extent of goal alignment (see Chapter 5). However, they note that the present construction market makes the use of the 'power approach' rather unlikely. They then, tentatively, propose the incentive of future work as *the* means to curb opportunism and facilitate a stable cooperative relationship. In his later writings, Cox (1999, 2001, 2004a, 2004b) concentrates on his 'power approach' to facilitating cooperation in buyer–supplier relationships. This framework has also been applied in analysing buyer–supplier relationships in the construction industry (Cox and Ireland, 2002, 2006).

This argument on the role of repeated exchanges ('repeat games') in creating and maintaining stable cooperative relationships is based on Axelrod's (1984) 'tit for tat' strategy. Axelrod argues that rational egoists will cooperate if future transactions offer benefits that outweigh those available by short-term opportunistic behaviours. On the face of it, such an argument is devoid of the existence of norms and so fits well with Cox's framework. However, for Cox's proposition to work there needs to be some shared expectation(s) about the future (i.e. *trust*). Further, this shared expectation about the future is created, largely, on the basis of past and present actions, *not* vice versa (Oberschall and Leifer, 1986; Zucker, 1986; Bradach and Eccles, 1989; McKnight *et al.*, 1998). Therefore, Cox's explanation for cooperation is tenable once trust exists. The argument that single-game encounters between contractors and one-off procurers of construction (i.e. clients) *always* lead to opportunistic bargaining/positioning is rather simplistic (Cox and Ireland, 2006). In construction, the communities of practice ensure that contractors are, to a reasonable degree, self-regulating. Retaliatory and reputational sanctions (in the market place) are very strong and pervasive when a contractor short-changes a client in a construction contract (Winch, 2001).

Cox's 'power approach' has been greatly influenced by Emerson's (1962) power-dependence theory. According to Emerson, an actor's social 'power resides implicitly in the other's dependency' (1962:32). The dependency is generated when one actor's access to a valued resource is mediated by another actor. Any analysis of power relations, therefore, reduces to analyses of dependencies. Emerson refers to such power as 'structural' in the sense that it reflects the properties of the organisational relationships with no regard for the social psychology of the individuals within the organisations. Yet, Emerson provides an all-inclusive definition for resources, ranging from material resources to organisational commitment and status, and leaves the operationalisation of this theoretical concept open. There are two problems with such a conceptualisation. Firstly, organisational commitment and status are attributes of the social psychology of individuals. Secondly, applying an 'exchange theory' perspective to all social relations is an undertaking that finds little favour with social identity theorists (cf. Tajfel, 1978, 1982; Tajfel and Turner, 1979; Turner, 1981, 1987). Such an approach proffers a notion of resources that is so ubiquitous in nature that it renders the theory of instrumentality of little utility (Montada, 1996; Tyler and Blader, 2000).

Emerson's conceptualisation of power assumes organisations to be in competition for control over some critical resources. In this regard unequal power relationships are unstable since they encourage the use of power, which in turn invokes responses in the form of 'cost reduction' and 'balancing operations'. Emerson noted that these reactions to imbalance in relations were akin to the idea of distributive justice. In short, dependency is bad and will (or *should*) be avoided. Emerson did not specify the conditions under which a power advantage will be used, suggesting this as the subject of future studies. However, this is a simple problem for (neo)classical economists. In classical and neoclassical economic theory, social exchange is primarily economic in nature, involving an exchange of goods, services and/or money. Also, the rational economic agent will *always* make full use of any power advantage in an exchange relationship. In fact, Cox and Ireland (2006) posit power as the principle for value appropriation in a dyadic exchange relationship. This conceptualisation reinforces the existing stereotype of power as a corrupting phenomenon in relationships. It is thus surprising that Cox and Ireland (2006) propose buyer dominance as a suitable power regime for developing stable cooperative buyer–supplier relationships. The three main sources of structural power are resource control, hierarchical authority and network centrality (Emerson, 1962; Astley and Sachdeva, 1984). These three sources of power interact with one another. Thus, given the context of a construction project organisation, industry practice, for example 'professional governance' and 'trilateral governance' (Winch, 2001), means that a client's unilateral power to act opportunistically is diluted.

However, power can also be used to influence others to reach cooperatively linked organisational goals (McClelland, 1975). This positive view of power underscores Liu and colleagues' 'power paradigm' in project management (Fellows *et al*, 2003; Liu *et al.*, 2003; Liu and Fang, 2006). In this regard, Tjosvold (1984) argues that power should be defined independently of goal interdependence. This is particularly important in the context of construction since the parties in a project are selected on the basis of, or at the very least, expectation of complementarity. Therefore, the respective parties are not in competition *per se*. This should pose problems for anyone attempting to apply Cox's 'power approach' to construction projects. This is not to suggest that the goals of parties in a construction project are optimally aligned. Even *within* organisations, there exists considerable scope for sub-optimisation of organisational goals. There is still the challenge of combating attachment to sub-goals in order create identification with the cooperatively linked (project) goals (Simon, 1991; Tyler and Blader, 2000). Reviews of industry practices suggest that construction clients' approach to expressing power has largely been negative (e.g. Green, 1998, 1999; Murray and Langford, 2003). However, this does not erase the conceptual distinction between these two uses of power. Thus, it is important to decouple the definition of power from its uses not least because the theorising of power influences practice. Indeed, Cox's writings target practitioners.

Price, authority and trust

Therefore, while somewhat different, Cox's power approach, like Williamson's TCE approach, is grounded in exchange theory. This has as its primary focus, *wealth maximisation* but given bounded rationality, may more practically be expressed as *satisficing*. In the context of construction projects, the TCE framework implies that the choice of appropriate governance structure for a contractual relationship is a transaction cost issue. However, researchers have been eager to develop and/or promote typologies of contracting strategies rather than explaining (sub)contracting *per se*. These typologies assume market and hierarchy as defining the opposite poles of a continuum of relational contracting strategies. In this way a typology of contracting strategies is thus constructed, ranging from direct market competition to strategic alliances. Prescriptive rules are then formulated to guide the selection of the 'fit for purpose' contracting strategy. For example, Cox and colleagues (Cox and Thompson, 1997; Thompson *et al.*, 1998) used their 'relational competence analysis' approach to formulate a typology of optimal contracting strategies and supply relationships (Figure 4.1). Many other researchers have developed similar typologies (e.g. Krippaehne *et al.*, 1992; Winch, 2001). These typologies generally have an intuitive and logical appeal and are followed in one form or another by construction companies.

However, the methodological basis and assumptions of these typologies are less sound and not borne out in practice. Firstly, it is not clear that wealth maximisation is always *the* goal. As Simon (1991) argues, even at the top executive levels of firms, there is considerable opportunity for conflict to exist between the goal of ownership (profit) and the goals of managers. Firms are represented in transactions by their managers who may not be owners at all, or only minor shareholders. Individuals' self-concepts are shaped by their organisational relationships (Tyler and Blader, 2000). In other words, multiple goals exist. Also what constitutes costs and benefits is not always clear-cut. Many people derive satisfaction from participating in the 'costly' activities undertaken in pursuit of ostensible organisational goals. Trust development research shows that a lot of effort is required to construct an

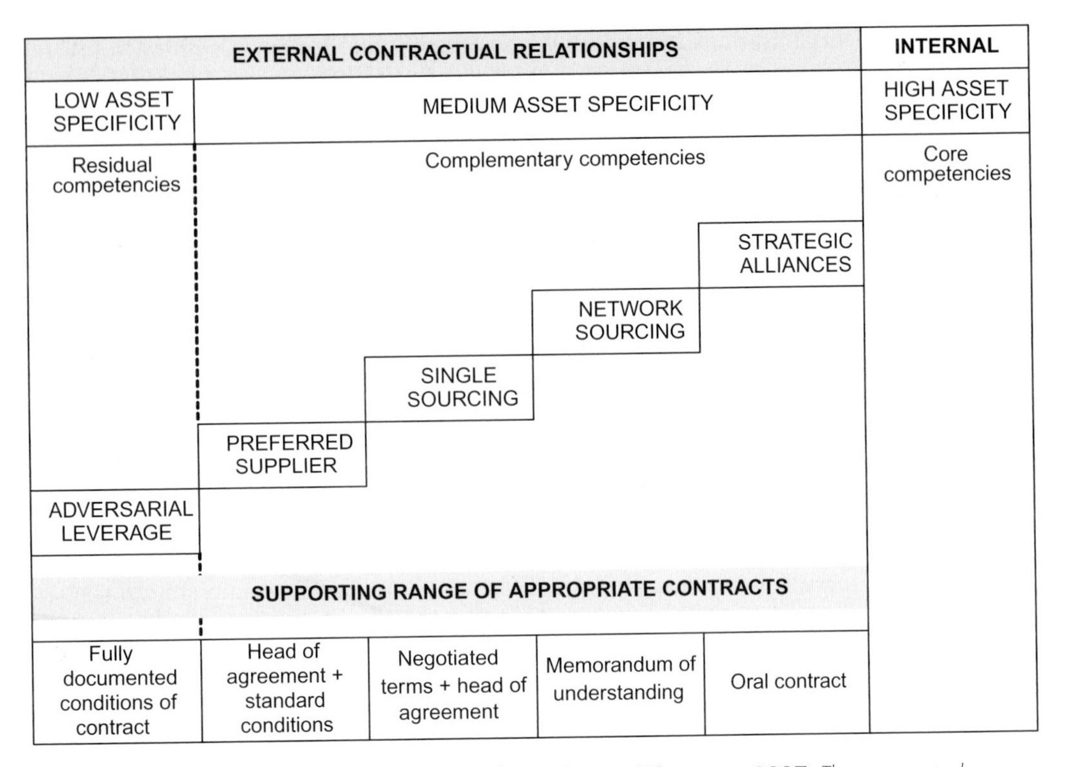

Figure 4.1 Optimal contracting model (adapted from Cox and Thompson, 1997; Thompson *et al.*, 1998)

'indefinite transaction future' to curb opportunistic behaviour (Ring and Van de Ven, 1994; Das and Teng, 1998; Jones and George, 1998; Rousseau *et al.*, 1998). Asset specificity and uncertainty can grow out of this trust-building effort rather than pre-existing the relationship. Thus, over time what constitutes costs and benefits may be ambiguous (Oberschall and Leifer, 1986). This poses problems with operationalising the TCE framework in contexts like construction which are people-centred. Turner and Simister (2001) note that a project's out-turn cost is not independent of the contractual framework (i.e. conditions of contract and pricing mechanisms). This is because the different contractual frameworks have different motivational influences on a contractor and, so, will lead to different out-turn costs. An ordering of contractual frameworks on the efficiency criterion (i.e. transaction costs) is problematic when the framework for the ordering is itself derived from the contractual framework in place. Further, Turner and Simister (2001) observe that the differences in out-turn cost of using different contractual frameworks are greater than the differences in transaction costs.

Secondly, in construction, as in other project-based industries where products are manufactured to specifications, contracts require a massive exchange of information *across* organisational boundaries in both negotiation and execution. The levels of communications are, generally, comparable to those observed between departments of a firm. In this regard, Winch (2002) conceptualised project organisations as information-processing systems. This heightens contracting uncertainty, beyond that due to price uncertainty (Winch, 1989). Under the TCE framework, this situation is better governed under a hierarchy (i.e. through vertical integration). But the pervasive use of *subcontracting* in the construction industry demonstrates that such information exchange is quite feasible and does not require (or necessarily lead to) vertical integration. This blurs the distinction between market communications and internal communications and, hence, renders vague the TCE criteria for choosing between them (cf. Simon, 1991).

Thirdly, there is evidence that price, authority and trust are woven together to govern transactions in construction. These provisions are contained in, or routinely introduced as special conditions in, standard forms of contract. This evidence is alluded to by TCE researchers in construction (e.g. Winch, 1989; Cox and Thompson, 1997; Pietroforte, 1997; Lai, 2000). Price competition is a major feature of the selection methodologies of most construction clients. Even where contracts are largely let on the basis of non-price criteria, there is usually a requirement to demonstrate price benchmarking. Some standard forms of contract contain express terms that require contracting parties to act in a spirit of mutual trust and cooperation (e.g. Institution of Civil Engineers, 1993; Association of Consultant Architects, 2000; Collaborating for the Built Environment, 2003). These standard forms of contract are, essentially, hierarchical documents (Stinchcombe, 1985; Thompson *et al.*, 1998):

- A method for adjusting costs, prices, quantities and payments.
- A legitimate authority (the *engineer*).
- Agreements about how, when, and who has authority, to modify contractual provisions, resolve disputes.

Winch (2001) alludes to the institution-based trust construction contracts generate through the use of 'professional governance', 'trilateral governance' and industry 'communities of practice' and, also, the cognition-based trust generated by 'reputation trading' in selection methodologies (cf. Zucker, 1986; McKnight *et al.*, 1998; see Chapter 6). There are many examples of long-term and stable (trust-based) relationships between main- and sub-contractors that periodically test the market by soliciting rival bids from other sub-contractors (Eccles, 1981; Hampson and Kwok, 1997; Kale and Arditi, 2001). There are also numerous examples of term contract arrangements between client and main contractor that are established, at least initially, through competitive tendering. Therefore, attempts to pitch market, hierarchy and relational contracting, or their operational derivatives – price, authority and trust respectively – either as mutually exclusive governance mechanisms or alternatively as defining a continuum are misleading. As Bradach and Eccles (1989) demonstrate, price, authority and trust are interdependent coordination mechanisms that can be combined in a variety of different ways.

Macneil's relational contract theory

The foregoing discussion has shown that (sub)contracting is not explained by the TCE framework. Winch (1989) acknowledges this when he describes contracting as a market failure, but blames this on the political preferences of powerful interest groups. Based on the foregoing, the authors conclude that the TCE framework is an inadequate guiding theory for contracting. So, is there a suitable theory that underpins contracting? The authors' answer, in the affirmative, is Macneil's (e.g. Macneil, 1974; 1978; 1985; 2000) *relational contract theory*. Macneil spent a greater part of his career developing his relational theory of contract. In fact, like Emerson, Macneil sees a much greater role for his theory – as a social theory of exchange. Of course, this has been criticised (Foster, 1982; Campbell, 1990). As stated earlier, the authors limit their analysis to economic exchange. It is impossible, and certainly not the aim of this chapter, to review all of Macneil's writings. Fortunately, the many examples of relational contracts that Macneil devoted a great deal of his writings to elaborating are familiar, or easily accessible, to many researchers in the construction industry. This review is, therefore, limited to the pillars/foundations of Macneil's relational theory of contract. Whitford (1985) and Campbell (2001) provide admiring reviews of Macneil's work. Empirical uses (and misuses) of Macneil's work also abound.

Macneil defines a contract as a process of projecting (economic) exchange into the future. Contracts become necessary wherever there is division of labour and specialisation. As all human action is constituted within social relations, contracts are instruments of social cooperation. While the (competing) self-interests of parties motivate them to enter into contractual relations, this competition must be bounded by an integral acceptance of cooperation. Macneil, as also observed by Winch (1989), emphasises that the problem of sustaining cooperation *within* contracts is more central than the threat of externalities. The boundaries of such legitimate negotiation and competition are set by the ten common contract behavioural patterns and norms (Macneil, 2000):

- *Role integrity* – roles require consistency, involve conflict and are inherently complex (also see Simon, 1991).
- *Reciprocity* – Macneil emphasises that this does not imply equality but 'some kind of evenness', what Ring and Van de Ven (1994) call 'fair dealing'.

- *Implementation of planning* – determining when liability begins, specifying contract terms and setting the remedy upon breach.
- *Effectuation of consent.*
- *Flexibility.*
- *Contractual solidarity* – the word 'solidarity' generally implying trust and confidence.
- *The restitution, reliance and expectation interests (the 'linking norms').*
- *Creation and restraint of power (the 'power norm').*
- *Propriety of means.*
- *Harmonization within the social matrix* – 'social matrix' implying supracontract norms.

Macneil maintains that this list of common contract behavioural patterns and norms is not exhaustive. These *internal* norms link to *external* norms, including sovereign law, industry customs and rules of trade associations and professional organisations. It is not intended in this chapter to elaborate on these intra- and supracontract norms. The point is to explain that there is a normative context in relation to contract. These norms (both internal and external) define the boundaries for social behaviour: what is right, adequate, acceptable and just. Without such norms, Macneil argues, there is no exchange, but war. There is, therefore, contrary to the postulations of rational choice theory, considerable conformity in choice (see Oberschall and Leifer, 1986). In this sense, Macneil argues that all (economic) exchange is relational.

Complex contracts

The most widely referenced aspect of Macneil's work in construction management research, as in other fields (Campbell, 2001), is the metaphor of the spectrum of contracts. It simply tries to draw particular attention to a class of contracts that:

- Involve 'asset specific' or 'idiosyncratic' (i.e. non-transferable) investments.
- Are too complex (i.e. *ex ante* specification of terms is impossible, of long duration, and/or under uncertain circumstances) to be fully specified during pre-contract negotiations.
- Require the parties to adjust both their obligations and expectations *ex post* during and at the conclusion of performance.

This certainly will pass as a textbook description of construction contracting. Macneil argues that for contracts of this nature the long-run selfish interests of the parties are better served by adopting cooperative attitudes and by preserving the contractual relationship. In other words, the pursuit of long-run self-interest requires the subordination of short-term self-interest. Such contracts cannot be governed efficiently by (neo)classical contract law, since that law emphasises strict contractual compliance, liabilities and remedies. Measurement problems, arising from the inherent complexity, complicate the determination and apportionment of liabilities, thus creating adversarial relationships. Paradoxically, research shows that this is also true of neoclassical-contract-governed incentivisation schemes (Ashley and Workman, 1986; Thompson *et al.*, 1998). Macneil called this special class of contracts *relational contracts*. Recognising the confusing terminology (where 'relational' applies to *all* contracts and at the same time to *a special class* of contracts), Macneil (2000) subsequently referred to this special class of contracts as 'complex' or 'intertwined'. The word 'complex' is preferred as it will resonate with many researchers and practitioners in construction. The explication of this class of contracts naturally required Macneil to contrast them with their conceptual opposites, 'discrete' contracts, not that these exist in reality. To solve this confusion in terminology, Macneil later used the term 'as-if-discrete' for his classificatory analyses. Therefore, discreteness and relationality are relative phenomena, depending on the differing emphasis placed on the common contract behavioural patterns and norms. Intensification of the contract norms of a competitive character, for example, implementation of planning and effectuation of consent, that is, attempts to closely specify and impose strict liabilities, will lead to contracts with discrete patterns of transactions (i.e. 'transactionised' contracts). These forms of contract may thus be suitable for contracts with strong discrete elements (e.g. stock dealing), whereas the contract norms of role integrity, preservation of the relation (i.e. trust), flexibility and harmonisation of relational conflict are very important in complex contracts.

Macneil finds negligible differences between classical and neoclassical micro-economic analyses of exchange. Both are based on the assumption of discrete rational utility maximising behaviour of economic agents. The only difference is that neoclassical economics concedes that relations underpin all economic exchange, but then typically explains away their impact with *Ceteris paribus*. This non-contextualisation of economic analyses invariably leads to fallacious findings (i.e. very remote from reality). Correspondingly, there is non-use in business of the formal classical contract remedies (Macaulay, 1963). Neoclassical contract law attempts to address the inaccuracy or indeterminacy of classical contract law, for example by drawing on industry customs in adjudicating disputes, and use of neoclassical ideas such as 'unconscionability'. Some non-common law jurisdictions also require the exercise of 'good faith' as a baseline obligation. As these are applied within classical contract principles, they are stopgap measures and may be counterproductive (Campbell, 2001). Thus, the main point of Macneil's discrete/relational spectrum is to stress that complex contracts be treated under a different form of analysis: relational analysis. But since the majority of business exchanges involve a web of relations rather than discrete transactions, Macneil makes a case for the conceptual validity and explanatory economy in, but does not argue for, having a relational theory for *all* contracts, not least because this will considerably pare down the list of exceptions in the neoclassical law of contract (Campbell, 2001). However, the main argument of the authors is that an application of relational analysis to commercial construction contracting should be an absolute necessity (Feinman, 2000).

Construction contracting

Macneil's work has influenced the new institutional economics, which seeks to identify transaction cost economising institutions. Working in this field, Williamson (1985, 1996) used Macneil's metaphor of the spectrum of contract to develop his TCE framework. As can be inferred from the earlier discussion, Williamson's TCE framework seems to have dominated construction management research. The new institutional economics is different from neoclassical economics in that it accepts, rather than explains away, that complex ('relational') contracts constitute a web of relations, but this difference blurs in the explication of how such relationships evolve and are governed. The assumption of individual utility maximisation and the 'denial' of norm-referenced behaviours (through game-theoretical modelling of cooperation) are features of neoclassical microeconomic analysis. This 'denial' of the operation of norms is sometimes concealed in notions of 'repeat trading' and/or 'long-term supply contracts' as being the paradigms of relational contract (Macneil, 2000). TCE discourses in construction management research are replete with this latter form of argument. To be adequate, the technical function of transaction cost analysis needs to be subsumed to, and understood in terms of, the relations the costs represent. Campbell (2001) provides an extreme example, which puts this argument in perspective: if *all* the costs of exchanging were to be reduced to zero, no exchange would take place. Therefore, the notion of the TCE approach as the basis for understanding the institutions of economic action is very remote (Campbell, 2001; cf. also Oberschall and Leifer, 1986; Turner and Simister, 2001). Thus far, the following can be deduced about the features of construction contracting from the discussions (see also Whitford, 1985):

- Standard form contracts are essentially 'incomplete', requiring future adjustment processes of an administrative kind.
- The wealth maximisation concerns of each party in the exchange are best served through cooperation and the subordination of short-term self-interest.
- As whole person relations form an integral and central aspect of contractual relationships, the social psychologies of the persons involved are very important. Thus, wealth maximisation concerns are not the only reason for the preservation of relations: intrinsic motivation and status concerns are but two other goals (cf. Ring and Van de Ven, 1994; Tyler and Blader, 2000).

- Price, authority and trust (among others) are interdependent contract norms that can be combined in different ways.

Recognition of the web of interdependent relations and the need for cooperation in construction contracting has, however, been slow and incremental, yet very significant. The transactionising tendencies of competitive tendering, resulting from the wave of privatisation of public services and the consequent breakdown of cooperative norms, are now being addressed through relational techniques in contract award and performance. These include, among others:

- The use of approved lists and knowledge of parties' reputation and past dealings.
- The change in government policy favouring more flexible and less prescriptive regulation (e.g. best value in the UK).
- The move from the use of technical specifications to performance specifications.
- Promoting the use of partnering (and alliancing), mechanisms for trust and team building (cf. Anvuur and Kumaraswamy, 2007), including use of voluntary non-adjudicatory dispute resolution methods with legislative support.
- Drafting of new standard forms of contract based on fairness principles and facilitative of cooperative attitudes, and the introduction of statutory legislation and/or guidelines to the same effect as supra-contract norms.
- Promoting more interaction among parties through the use of more integrative (*than* design–bid–build) methods of procurement.

These changes are, it is suggested, part of the move towards 'cultural transformation' of the construction industry, which has often been misrepresented and misunderstood. While the current neoclassical contract law may not adequately track these cooperative relationships, it is important to recognise these developments as constrained efforts in the right direction. Researchers in construction management have a significant role to play in this, not least because their writings have a significant impact on practitioners. As a fledgling field of enquiry, construction management research has no formal theories of its own, often relying on *ad hoc* classification systems, taxonomies and conceptual frameworks. However, its further development – to theoretical systems – is contingent on how much attention is given to resolving/clarifying methodological issues. For example, many researchers while citing Macneil and Williamson's works either have glossed over the differences in their philosophical bases or have assumed these bases to be the same. Also, many researching relationship marketing that cite the IMP Group's interaction model also cite Gummesson (1997, 2002). This would link both approaches to Williamson's TCE approach. However, careful reading of Gummesson (1997) shows that this approach has conceptual underpinnings similar to Macneil's relational contract theory:

- All exchange is relational.
- Fair dealing, trust and its correlates, as exchange norms.
- Harmonisation of competition within a cooperative institutional/ regulatory framework.
- 'There is not a single point in the cause of a relationship when a sale is made . . . the sale has never been fully consummated until the project has been completed to the client's satisfaction.' (Wittreich, 1969:9)

Clearly, superficial research may result in misleading research findings and/or may complicate efforts to relate and integrate the contributions of other researchers both within and across research disciplines. Equally, to reject the use/influence of management strategies, like partnering, on the ground that they are 'outside' the neoclassical law of contract, or to conceptualise/propose strategies to transactionise construction contracting because of the prevailing adversarial construction climate (i.e. lack of trust), is intellectually non-stimulating.

To the extent that cooperative behavioural patterns and norms are of central importance in construction contracting, it is reasonable to expect that the selected procurement method and standard form of contract should be facilitative of the development of these norms (Turner and Simister, 2001). Equally, relational analysis using the common contract behavioural patterns and norms might usefully be undertaken to determine the extent to which the contract delivery and governance structures are supportive of cooperative contract norms. The following section presents relevant findings from one of a series of longitudinal case studies undertaken in pursuit of this very objective. Case studies are especially important in providing fundamental understanding of the structures and processes involved in construction projects and the decision criteria that underpin their use (Eisenhardt, 1989; Yin 1994; Gummesson, 2000).

Case study methodology and methods

The research aim is, *inter alia*, developing a management support system to aid the formulation of more effective and efficient construction procurement and operational systems (Kumaraswamy *et al.*, 2006). The case studies were all selected to suit the demands of a theoretical sampling frame (Glaser and Strauss, 1967). The case studies seek an in-depth understanding (and hence, explanation) of the selection practices/

processes of construction clients, the frameworks/paradigms that guide those processes, their influences on the cooperative behaviours of project team members and team members' rationalisations of these processes and their impacts. The case studies also seek to identify and codify any generalisable knowledge ('best practices') that can be incorporated into management support systems. The case studies thus seek both generalisation and particularisation. An ethnographic approach was thus considered the most suitable. This approach enables a holistic view of, and understanding of, the structures, processes and driving forces that are the focus of this research project (Gummesson, 2000).

The selected case study was started in September 2004 and has reached close out. The research involved the study of existing project documentation, minutes of project meetings, official correspondences, reports and press commentaries. Systematic observation of project meetings, in-depth semi-structured interviews with managerial personnel and short structured questionnaire surveys were started in April 2005. The use of different methods to collect both qualitative and quantitative data provides a triangulated research framework and, hence, better assurance of the reliability of the results (Eisenhardt, 1989; Gummesson, 2000; Yin, 2003). A cross-section of meetings observed included the directors' review meetings, design meetings, M&E coordination meetings, procurement (of subcontract works packages) meetings and adjudication meetings. Managers and executive directors represented the various functions (i.e. firms) at these meetings. Debriefing notes were made to capture observations of interactions and seemingly critical issues that should be investigated in the in-depth interviews.

A total of 21 project personnel were interviewed, comprising six executive directors, six senior managers and nine managers from the client, consultants, main and sub-contractors. Some of the senior managers were interviewed twice. The sub-contractor personnel interviewed were a senior manager from the building services sub-contractor and the executive director and CEO of the structural steel sub-contractor. All the interviews generally followed a common interview protocol, which included questions on the characteristics and objectives of the project, project governance structures and processes, project performance criteria, measures and outcomes, and team-building practices. The interviewees also completed short questionnaires designed to measure their perceptions of interdependence, fairness of decision-making procedures, processes and outcomes, and intrinsic job satisfaction. The six executive directors were also asked to state their company's corporate objectives and then to rate the importance of these objectives, and others (not mentioned by them) in a set of ten commonly cited corporate objectives presented to them, on a Likert scale (0 = not all important; 10 = extremely important). Issues that were also identified as needing further probing from the project documentation and observations were taken up with the respective managers and/or directors who had an input into, or were affected by such issues.

Constraints on time and resources did not allow extensive direct observation of decision-making processes on this project therefore a critical incident technique was applied (Flanagan, 1954; Andersson and Nilsson, 1964) to ensure that all potentially critical situations were captured. The respondents were asked to think of the most difficult problem encountered on this project for which the nature and timeliness of their response was 'make or break' for the project. They were asked to describe the circumstances of the incident, what specific actions they took to resolve the problem and how they perceived the responses of the other project actors. The interviews with the managers each lasted about 1 hour and those with senior managers and executive directors took 1–2 hours. Each interview was recorded and transcribed, producing over 280 pages that were coded to reflect common themes and other emergent issues.

This intensive ethnographic approach was time-consuming and sometimes cumbersome, yet produced a rich range of data. Applying the critical incident technique proved particularly useful: remarkably, all the respondents identified the same incident – a major structural redesign – as the most difficult challenge encountered on the project. However, descriptions of the circumstances of the incident varied, senior managers and executive directors displaying, and expectedly so, greater knowledge of the background to the problem than other managers. (It was during this incident that attempts to gain access to the project were unsuccessful, thus, the critical incident technique helped fill the time gap.) The next section describes the case study project and discusses relevant findings.

The project: nature, scope, objectives and strategies

The client in this project, a prestigious property developer, owns and manages prime office and retail space in Hong Kong. The project involved the redevelopment of an existing office tower in Hong Kong at a total construction cost of over US$80 million. The project objectives were: to create a small grade 'A' office building; to complement and extend an existing luxury retail space; and also to provide more food and beverage space. Client drivers included the need to provide assurance that construction costs were competitive and reflected the current market price levels and to improve relationships with contracting partners through a partnering 'offensive'. The scope of works involved demolishing the existing office tower, supplementing the existing foundations and erecting a steel-cored superstructure tower (25 floors). This involved three separate and sequential contracts for the demolition, foundation and the superstructure works. Table 4.1 provides summarised details of the procurement and contractual arrangements in this

Table 4.1 Project procurement and contractual systems

Contract system	Demolition	Foundation	Superstructure
Procurement arrangement	■ Modified management contracting; no direct works by main contractor; all three sub-contract work packages for demolition, service diversion/termination and provision of hoardings and protective deckings tendered competitively	■ Traditional approach with no sub-contract portions	■ A form of management contracting; named* sub-contract packages for structural steel, curtain walls, lifts and escalators and building services; 90 domestic and 22 provisional sum work packages progressively procured through competitive tendering
Contractual measures that support arm's-length relations	■ Traditional JCT form of building contract, with retention fund and performance bond ■ Negotiated lump sum fixed price and fixed duration contract	■ Traditional JCT form of building contract, with retention fund and performance bond ■ Negotiated fixed price lump sum contract	■ Traditional JCT form of building contract, with retention fund, performance bond and liquidated damages associated with four project milestones ■ Indemnify client against all delay/disruption claims arising from or in connection with the foundation contract
Contractual measures that support cooperative relations	■ Nil	■ Lump sum 'win or lose' on-time completion bonus ■ Compensation for delay in services diversion in connection with demolition contract	■ Negotiated GMP contract with sharing of savings from value engineering, procurement and the design development allowance, 60/40% between client and main contractor and, overruns at 100% to main contractor

* See text

project. All three main contracts were negotiated and let to the same main contractor under the JCT-based HKSAR standard form of building contract (private edition – with quantities). Formal partnering with a non-binding charter was adopted with the active involvement of the client, consultants, main and subcontractors.

Project complexity dimensions

The client required sectional completions to facilitate early hand-over of lower floors of the building to tenants to fit out as the works progressed. Three milestones, each related to the receipt of a temporary occupation permit (TOP), were established. Achieving these key dates was a priority for the client, communicated clearly to everyone involved and reinforced with liquidated damages clauses. The site is small and adjoins existing luxury retail and hotel space. The process for obtaining statutory approvals was very cumbersome, as some of the government departments concerned do not have specified turn-around times for dealing with submissions. The foundation works required that a new basement be constructed within an existing basement with an 8 m head of water. When the existing office tower was built, the adjoining street collapsed into the site and there was a risk of history repeating itself. The final method and sequence of basement construction adopted was innovative within Hong Kong. The superstructure steel core was also the first of its kind for the client and was off-centred so that many of the floors were cantilevered. These factors introduced a steep learning curve, reduced flexibility in methods of and access for work and transportation of materials, increased site safety risks and created coordination problems. An extensive structural redesign, made 2 months into the construction stage, coupled with the large numbers of tenant-induced change requests, exacerbated the coordination problems. These factors interacted and transformed the project into a complex and risky undertaking.

Selecting a partner

The client entered into single source negotiations with the main contractor for three main reasons. First, the main contractor has a reputation of being, perhaps, the best piling contractor in Hong Kong, and was quite familiar with the ground conditions in that area. Second, the main contractor and the client belong to the same holding company, giving the client some significant corporate leverage. Third, the main contractor is committed to cooperation, adopting partnering as a corporate strategy on all projects, and has a history of successful past dealings with the client on similar projects. However, demonstrating competitiveness of the main contractor's pricing in the market was still a major requirement for award. All the consultants consistently stated as part of their project goals, the need to project and maintain a good long-term relationship with the client. They all (except the M&E consultant) had a previous history of successful dealings with the client on one or more projects involving the guaranteed maximum price (GMP) methodology. The client, main contractor and consultants were, therefore, reasonably familiar with the 'rules of the game' (e.g. negotiations, GMP methodology) and had built up some trust while working on an open-book basis.

Project overview and performance

Many of the problems encountered carried very high risks for this project, but all of the problems were resolved and have been turned into major successes for the project team. A shared sense of interdependence, respect and trust was evident amongst the team. This was reflected in the quantity and quality of project communications, problem-solving, prosocial behaviours and, consequently, in the project performance outcomes.

The foundation contract was completed ahead of schedule. All the TOPs for the superstructure contract have been consistently achieved. The occupation permit (OP) was originally expected at the end of October 2006. However, due to a statutory requirement in respect of the approval of some parts of the work, some of which was done 40 years ago, the OP has been phased. The first-phase OP, involving the basement and eight floors, was achieved as scheduled and that for the rest of the building was granted at the end of November 2006. The out-turn quality of the works was adjudged very good despite some disappointment with the quality of some of the early designs and the coordination of design changes. No serious site accident was registered and no complaint of major disruption to adjoining businesses was lodged. The adjudication process has been largely devoid of any serious disputes. The few remaining disagreements, mostly to do with M&E design changes, are not expected to go beyond senior management level. With final accounts settlement due in January 2007, the out-turn cost is expected to exceed the budgeted cost by 1%. Some significant but modest procurement savings were derived (i.e. cost savings resulting from the letting of work packages for less than the budget allocated to them in the GMP contract). However, these were transferred into the design development fund, which is expected to be fully spent. Therefore, there are no savings to share.

The project has won six industry awards and is expected to achieve the (highest) platinum rating in the prestigious Hong Kong Building Environmental Assessment Method (HK-BEAM) certification scheme. Yet, it seemed that the project's performance outcomes presented an incomplete picture of the level and impact of the teamwork experienced on the project.

Incentives, motivation and cooperation

Incentive and sanctioning systems

The traditional form of contract supports arm's-length contractual relations that only lead to contractual compliance (Thompson *et al.*, 1998). The fixed-price lump sum contracts create a conflict of motive between the client and main contractor (Bower *et al.*, 2002). Measures that allow some sharing of project risks and the associated pain/gain are considered as supportive of cooperative relations (Bower *et al.*, 2002; Thompson *et al.*, 1998). The foundation and superstructure contracts had a fair number of such measures, as indicated in Table 4.1. However, several factors associated with the intent, design, operation or outcome of these systems, raise doubts over what impact, if any, they have had on the cooperative behaviours of the project actors.

First, the liquidated damages were based on cost (i.e. overheads) and were not related to the probable loss in rental income. The client also demanded guarantees on both the maximum cost and time for the project. This can be a recipe for disputes. However, conscious and deliberate choices were made by the main contractor (including some sub-contractors), on the one hand, not to claim on the provisions in the conditions of contract and, by the client, on the other hand, not to utilise the contractual safeguards. There was very little scope to create cost savings. The designs were 90% complete and the main contractor was engaged for the most part of that process. As such, all the value engineering was done pre-contract and the savings incorporated into the GMP. Post-contract, the structural redesign effectively removed any scope (i.e. available float) for design refinement and value engineering. About 60% of the value of the works fell under *named* domestic sub-contractors. These are, essentially, nominated sub-contractors, except that in this arrangement the client is not legally liable for their default. The design development allowance was very modest (less than 2% of the GMP). Added to this was the fact that the main contractor had no control over the change request (or architect's instructions) process.

Non-instrumental motivations

What then was the motivation for the high level of cooperation witnessed on this project? To start with, it is clear from the foregoing that the answer does not lie in the incentive and sanctioning systems used. There was consensus that the formal aspects of the partnering process – workshops, champions' meetings and periodic evaluation – were often neglected. Perhaps this finding is suggestive of conceptual–definitional problems associated with the entity of partnering (see Bresnen and Marshall, 2000). There was a dominant client culture underlying the decision-making processes in this project. For the client's in-house project management team, delivering projects through cooperation was a way of working that they enjoyed. The GMP contract, with the open book approach, provided the mechanism to cooperate by unifying the motives of the client, consultants and the main contractor (Bower *et al.*, 2002). A distinctive feature of this project was the high number, frequency and long duration of meetings and all agreed that these had very little to do with the procurement or contractual arrangements. One project director, who was new to the team, felt that in a sense the meetings had become self-sustaining and prevented people from 'just getting on with the job'. While there was a general tendency for the in-house project management team to micro-manage, the benefits of participation were obvious. These meetings, undertaken in pursuit of ostensible goals, did not involve costs alone. Clearly, some members derived satisfaction from participating in these 'costly' (in TCE terms) activities (see Oberschall and Leifer, 1986).

It appears, however, that the cooperation and good relationships on this project related more to the fairness of the client's decision-making processes and outcomes. The consultants did not tender on fees and their remuneration was fair and adequate. Interim payment certificates were honoured promptly (within 14 days). The change request system ensured that all subcontract variations were fully priced and agreed by the client, consultants, main and sub-contractors before they were implemented and any additional sums due were progressively included in the sub-contractors' interim certificates. The main contractor's profit margin and preliminaries were pre-agreed, fixed and then ring-fenced. The level of preliminaries was above the prevailing market rates. In addition, most of the high-risk sub-contract packages were included as provisional sums. This explains why the adjudication process has been essentially dispute-free. The following statement by an executive director from the main contractor company puts our arguments in perspective:

> *The pressure from us to claim is less and the consultants are not trying to defend their designs or to discredit our claims . . . Without these pressures, relationships are more harmonious. Yes, if we attack any other members (of the project organisation), we won't get any benefits. We just try to advise them their designs are not good, find out if we can still use the same design or may be find a cheaper one, then everybody is happy. Any benefits from this sort of design changes or alternative materials are shared. It's not from any one's pocket.*

Clearly, with nothing to 'fight' for, the best marketing for the consultants, main and sub-contractors was to do a good job, project and maintain a favourable image with the client in the hope of securing repeat business. This was held by all to be a major driver, as winning work is very expensive and difficult.

Concluding observations

The first part of this chapter argued that attempts to explain contracting in the construction industry using (neo)classical economics models, especially the TCE framework, further cloud understanding of the purpose of contracting and the significance of price, authority and trust as forms of control. It then argued that price, authority and trust are best considered as interdependent, but distinct, forms of control that can be combined in a variety of ways. The second part of this chapter used relevant findings from a case study project to explore the issues discussed in the first part. Site and design constraints, a commercial decision to target sectional completions and the changing business case for various tenant spaces transformed this project into a very complicated and high-risk endeavour. However, the project benefited from highly cooperative relationships and all project targets were met. It is clear from the foregoing that goal congruence and fairness motives, rather than the desire to win incentives and/or avoid sanctions, were the overriding determinants of the high level of cooperation experienced on this project.

The evidence also indicates that there is more to cooperative interaction than can be explained by transaction costs. The evidence generally supports Thompson *et al.*'s (1998) typology of contractual relationships, but also highlights the danger involved in constructing such rigid typologies. This project used a combination of arm's-length and cooperative contractual instruments. The disconfirming effect of the traditional form of contract on the cooperative behaviours of the team is particularly noteworthy. The presence of market prices helped this relationship by facilitating the development of trust. For this client, maintaining a very good relationship with its consultants, contractors and suppliers was sufficient to secure their cooperation in providing a high quality of service to the end-users. This also removed the 'hit or miss' element associated with achieving the stated project objectives. As Bradach and Eccles (1989:116) concluded, 'human reason and social circumstance lead to much more complex forms of control'.

References

Andersson, B-E. and Nilsson, S-G. (1964) Studies in the reliability and validity of the critical incident technique. *Journal of Applied Psychology*, 48(6), 398–403.

Anvuur, A.M. and Kumaraswamy, M.M. (2007) Conceptual model of partnering and alliancing. *Journal of Construction Engineering and Management*, 133(3), 225–234.

Aoki, M., Gustafsson, B. and Williamson, O.E. (eds.) (1990) *The Firm as a Nexus of Treaties*. Sage, London.

Ashley, D.B. and Workman, B.W. (1986) *Incentives in Construction Contracts*. A Report to the Construction Industry Institute (CII) Source Document 8, The University of Texas, Austin.

Association of Consultant Architects (2000) *PPC 2000: ACA Standard Form of Contract for Project Partnering*. Association of Consultant Architects (ACA), Bromley.

Astley, W.G. and Sachdeva, P.S. (1984) Structural sources of intraorganizational power: a theoretical synthesis. *The Academy of Management Review*, 9(1), 104–113.

Axelrod, R.M. (1984) *The Evolution of Cooperation*. Basic Books, New York.

Bower, D., Ashby, G., Gerald, K. and Smyk, W. (2002) Incentive mechanisms for project success. *Journal of Management in Engineering*, 18(1), 37–43.

Bradach, J.L. and Eccles, R.G. (1989) Price, authority, and trust: from ideal types to plural forms. *Annual Review of Sociology*, 15(1989), 97–118.

Bresnen, M. and Marshall, N. (2000) Partnering in construction: a critical review of issues, problems and dilemmas. *Construction Management and Economics*, 18(2), 229–237.

Bridge, A. and Tisdell, C. (2006) The determinants of the vertical boundaries of the construction firm: response. *Construction Management and Economics*, 24(3), 233–236.

Bridge, A.J. and Tisdell, C. (2004) The determinants of the vertical boundaries of the construction firm. *Construction Management and Economics*, 22(8), 807–825.

Campbell, D. (1990) The social theory of relational contract: Macneil as the modern Proudhon. *International Journal of the Sociology of Law*, 18(17), 75–95.

Campbell, D. (ed.) (2001) *The Relational Theory of Contract: selected works of Ian Macneil*. Sweet & Maxwell, London.

Campbell, N.C.G. (1985) An interaction approach to organizational buying behaviour. *Journal of Business Research*, 13(1), 35–48.

Chang, C-Y. (2006) The determinants of the vertical boundaries of the construction firm: comment. *Construction Management and Economics*, 24(3), 229–232.

Collaborating for the Built Environment (2003) *Be Collaborative Contract: multi-user CD-ROM*. Collaborating for the Built Environment (Be), Reading.

Cornes, D.L. (1996) The second edition of the new engineering contract. *International Construction Law Review*, 13(Part 1, January), 97–119.

Costantino, N., Pietroforte, R. and Hamill, P. (2001) Subcontracting in commercial and residential construction: an empirical investigation. *Construction Management and Economics*, 19(4), 439–447.

Cox, A. (1996) Relational competence and strategic procurement management: towards an entrepreneurial and contractual theory of the firm. *European Journal of Purchasing and Supply Management*, 2(1), 57–70.

Cox, A. (1999) Power, value and supply chain management. *Supply Chain Management: An International Journal*, 4(4), 167–175.

Cox, A. (2001) Managing with power: strategies for improving value appropriation from supply relationships. *Journal of Supply Chain Management*, 37(2), 42–47.

Cox, A. (2004a) The art of the possible: relationship management in power regimes and supply chains. *Supply Chain Management: An International Journal*, 9(5), 346–356.

Cox, A. (2004b) Business relationship alignment: on the commensurability of value capture and mutuality in buyer and supplier exchange. *Supply Chain Management: An International Journal*, 9(5), 410–420.

Cox, A. and Ireland, P. (2002) Managing construction supply chains: the common sense approach. *Engineering Construction and Architectural Management*, 9(5–6), 409–418.

Cox, A. and Ireland, P. (2006) Relationship management theories and tools in project procurement. In: Pryke, S.D. and Smyth, H.J. (eds.) *The Management of Complex Projects: a Relationship Approach*. Blackwell, Boston, pp. 251–281.

Cox, A. and Thompson, I. (1997) 'Fit for purpose' contractual relations: determining a theoretical framework for construction projects. *European Journal of Purchasing and Supply Management*, 3(3), 127–135.

Das, T.K. and Teng, B-S. (1998) Between trust and control: developing confidence in partner cooperation in alliances. *The Academy of Management Review*, 23(3), 491–512.

Dubois, A. and Gadde, L-E. (2002) The construction industry as a loosely coupled system: implications for productivity and innovation. *Construction Management and Economics*, 20(7), 621–631.

Eccles, R. (1981) The quasifirm in the construction industry. *Journal of Economic Behaviour and Organisation*, 2(4), 335–357.

Eisenhardt, K.M. (1989) Building theories from case study research. *The Academy of Management Review*, 14(4), 532–550.

Emerson, R.M. (1962) Power-dependence relations. *American Sociological Review*, 27(1), 31–40.

Feinman, J.M. (2000) Relational contract theory in context. *Northwestern University Law Review*, 94(3), 737–748.

Fellows, R., Liu, A. and Fong, C.M. (2003) Leadership style and power relations in quantity surveying in Hong Kong. *Construction Management and Economics*, 21(8), 809–818.

Flanagan, J.C. (1954) The critical incident technique. *Psychological Bulletin*, 51(4), 327–358.

Foster, K. (1982) Review of Ian Macneil: the new social contract. *Journal of Law and Society*, 9(1), 144–148.

Glaser, B.G. and Strauss, A.L. (1967) *The Discovery of Grounded Theory: Strategies for Qualitative Research*. Aldine, Chicago.

Green, S.D. (1998) The technocratic totalitarianism of construction process improvement: a critical perspective. *Engineering, Construction and Architectural Management*, 5(4), 376–386.

Green, S.D. (1999) Partnering: the propaganda of corporatism? *Journal of Construction Procurement*, 5(2), 177–186.

Gummesson, E. (1997) Relationship marketing as a paradigm shift: some conclusions from the 30R approach. *Management Decision*, 35(4), 267–672.

Gummesson, E. (2000) *Qualitative Methods in Management Research*, 2nd Edition. Sage, Thousand Oaks.

Gummesson, E. (2002) *Total Relationship Marketing: Marketing Strategy Moving from the 4Ps – Product, Price, Promotion, Place – of Traditional Marketing Management to the 30Rs – the Thirty Relationships – of a New Marketing Paradigm*, 2nd Edition. Butterworth-Heinemann, Oxford.

Hampson, K.D. and Kwok, T. (1997) Strategic alliances in building construction: a tender evaluation tool for the public sector. *Journal of Construction Procurement*, 3(1), 28–41.

Institution of Civil Engineers (1993) *The New Engineering Contract*, 1st Edition. Institution of Civil Engineers (ICE), Thomas Telford, London.

Jones, G.R. and George, J.M. (1998) The experience and evolution of trust: implications for cooperation and teamwork. *The Academy of Management Review*, 23(3), 531–546.

Kale, S. and Arditi, D. (2001) General contractors' relationships with subcontractors: a strategic asset. *Construction Management and Economics*, 19(5), 541–549.

Koskela, L. (2003) Is structural change the primary solution to the problems of construction? *Building Research and Information*, 31(2), 85–96.

Koskela, L. and Ballard, G. (2006) Should project management be based on theories of economics or production? *Building Research and Information*, 34(2), 154–163.

Krippaehne, R.C., McCullouch, B.G. and Vanegas, J.A. (1992) Vertical business integration strategies for construction. *Journal of Management in Engineering*, 8(2), 153–166.

Kumaraswamy, M.M., Palaneeswaran, E., Ng, T.S.T. and Rahman, M.M. (2006) Towards an integrated management support system for large clients. *Electronic Journal of Information Technology in Construction*, 11(Special Issue Decision Support Systems for Infrastructure Management), 197–210.

Lai, L.W.C. (2000) The Coasian market-firm dichotomy and subcontracting in the construction industry *Construction Management and Economics*, 18(3), 355–362.

Liu, A. and Fang, Z. (2006) A power-based leadership approach to project management. *Construction Management and Economics*, 24(5), 497–507.

Liu, A., Fellows, R. and Fang, Z. (2003) The power paradigm of project leadership. *Construction Management and Economics*, 21(8), 819–829.

Macaulay, S. (1963) Non-contractual relations in business: a preliminary study. *American Sociological Review*, 28(1), 55–67.

Macneil, I.R. (1974) The many futures of contracts. *Southern California Law Review*, 47(3), 691–816.

Macneil, I.R. (1978) Contracts: adjustment of long-term economic relations under neo-classical and relational contract laws. *Northwestern University Law Review*, 72(6), 854–965.

Macneil, I.R. (1985) Relational contracts: what we do and do not know. *Wisconsin Law Review*, 1985(3), 483–525.

Macneil, I.R. (2000) Relational contract theory: challenges and queries. *Northwestern University Law Review*, 94(3), 877–907.

Maylor, H. (2006) Special Issue on rethinking project management (EPSRC network 2004–2006). *International Journal of Project Management*, 24(8), 635–637.

McClelland, D.C. (1975) *Power: the Inner Experience*. Irvington, New York.

McKnight, D.H., Cummings, L.L. and Chervany, N.L. (1998) Initial trust formation in new organizational relationships. *The Academy of Management Review*, 23(3), 473–490.

Montada, L. (1996) Trade-offs between justice and self-interest. In: Montada, L. and Lerner, M.J. (eds.) *Current Societal Concerns about Justice*. Plenum Press, New York, pp. 259–275.

Murray, M. and Langford, D.A. (eds.) (2003) *Construction Reports 1944–98*. Blackwell Science, Oxford.

Oberschall, A. and Leifer, E.M. (1986) Efficiency and social institutions: uses and misuses of economic reasoning in sociology. *Annual Review of Sociology*, 12, 233–253.

Pietroforte, R. (1997) Communication and governance in the building process. *Construction Management and Economics*, 15(71).

Pryke, S.D. and Smyth, H.J. (2006) Scoping a relationship approach to the management of complex projects in theory and practice. In: Pryke, S.D. and Smyth, H.J. (eds.) *The Management of Complex Projects: a Relationship Approach*. Blackwell, Boston, pp. 21–45.

Ring, P.S. and Van de Ven, A.H. (1994) Developmental processes of cooperative interorganizational relationships. *The Academy of Management Review*, 19(1), 90–118.

Rousseau, D.M., Sitkin, S.B., Burt, R.S. and Camerer, C. (1998) Introduction to special topic forum: Not so different after all: a cross-discipline view of trust. *The Academy of Management Review*, 23(3), 393–404.

Shirazi, B., Langford, D.A. and Rowlinson, S.M. (1996) Organizational structures in the construction industry. *Construction Management and Economics*, 14(3), 199–212.

Simon, H.A. (1991) Organizations and markets. *The Journal of Economic Perspectives*, 5(2), 25–44.

Smyth, H.J. (2006) Measuring, developing and managing trust in relationships. In: Pryke, S.D. and Smyth, H.J. (eds.) *The Management of Complex Projects: a Relationship Approach*. Blackwell, Boston, pp. 97–120.

Stinchcombe, A. (1985) Contracts as hierarchical documents. In: Stinchcombe, A. and Heimer, C. (eds.) *Organisational Theory and Project Management*. Norwegian University Press, Bergen, pp. 121–171.

Tajfel, H. (1978) *Differentiation between Social Groups: Studies in the Social Psychology of Intergroup Relations*. Academic Press, London.

Tajfel, H. (1982) Social psychology of intergroup relations. *Annual Review of Psychology*, 33, 1–39.

Tajfel, H. and Turner, J.C. (1979) An integrative theory of intergroup conflict. In: Austin, W.G. and Worchel, S. (eds.) *The Social Psychology of Intergroup Relations*. Brooks/Cole, Monterey, pp. 33–47.

Thompson, I., Cox, A. and Anderson, L. (1998) Contracting strategies for the project environment. *European Journal of Purchasing and Supply Management*, 4(1), 31–41.

Tjosvold, D. (1984) Cooperation theory and organizations. *Human Relations*, 37(9), 743–767.

Turner, J.C. (1981) The experimental social psychology of *intergroup* behaviour. In: Turner, J.C. and Giles, H. (eds.) *Intergroup Behaviour*. Blackwell, Oxford, pp. 66–101.

Turner, J.C. (1987) *Rediscovering the Social Groups Self-categorization Theory*. Blackwell, Oxford.

Turner, J.R. and Keegan, A. (2001) Mechanisms of governance in the project-based organization: roles of the broker and steward. *European Management Journal*, 19(3), 254–267.

Turner, J.R. and Simister, S. (2001) Project contract management and a theory of organisation. *International Journal of Project Management*, 19(8), 457–464.

Tyler, T.R. and Blader, S.L. (2000) *Cooperation in Groups: Procedural Justice, Social Identity and Behavioural Engagement*. Psychology Press, Philadelphia.

Walker, A. and Chau, K.W. (1999) The relationship between construction project management theory and transaction cost economics. *Engineering, Construction and Architectural Management*, 6(2), 166–176.

Whitford, W.C. (1985) Ian Macneil's contribution to contracts scholarship. *Wisconsin Law Review*, 1985(3), 545–560.

Williamson, O.E. (1979) Transaction-cost economics: the governance of contractual relations. *Journal of Law and Economics*, 22(2), 233–261.

Williamson, O.E. (1985) *The Economic Institutions of Capitalism*. Free Press, New York.

Williamson, O.E. (1996) *The Mechanisms of Governance*. Oxford University Press, New York.

Winch, G.M. (1989) The construction firm and the construction project: a transaction cost approach. *Construction Management and Economics*, 7(4), 331–345.

Winch, G.M. (2001) Governing the project process: a conceptual framework. *Construction Management and Economics*, 19(8), 799–808.

Winch, G.M. (2002) *Managing Construction Projects: an Information Processing Approach*. Blackwell Science, Oxford.

Winch, G.M. (2006) Towards a theory of construction as production by projects. *Building Research and Information*, 34(2), 164–174.

Wittreich, W.J. (1969) *Selling – a Prerequisite to Success as a Professional*. Paper presented Detroit, Michigan, 8 January.

Yin, R.K. (1994) *Case Study Research: Design and Methods*, 2nd Edition. Sage Publications, Thousand Oaks.

Yin, R.K. (2003) *Case Study Research: Design and Methods*, 3rd Edition. Sage Publications, Thousand Oaks.

Zucker, L.G. (1986) Production of trust: institutional sources of economic structure. In: Straw, B.M. and Cummings, L.L. (eds.) *Research in Organizational Behaviour*. JAI Press, Greenwich, pp. 1840–1920.

Section II

Collaborative Relationships and Conceptual Frameworks

In this section collaborative relationships and conceptual frameworks within internal and external networks of relationships are explored. The section particularly builds upon **Chapter 4** at the end of **Section I**. The conceptual scope of collaboration within frameworks and projects constituted into programmes is developed. The behavioural concepts for performance improvement are conceptually explored.

In **Chapter 5** Aaron Anvuur and Mohan Kumaraswamy provide a conceptual appraisal of collaboration through cooperation and how improvements can be generated for theory and in practice. **Chapter 6** takes a further analytical view of developing trust. In this chapter Smyth reviews previous frameworks for conceiving trust. A reconsidered frame is provided, which takes greater account of factors of influence and power and issues of moral philosophy.

5 Better collaboration through cooperation

Aaron Anvuur and Mohan Kumaraswamy

Introduction

In construction, as in other project-based industries, the need for cooperation arises from uncertainty, interdependence and complexity (Shirazi *et al.*, 1996; Dubois and Gadde, 2002). Industry coping strategies designed to respond to these challenges have created a setting where commercial pressures, legal and contractual issues traditionally encourage reticence, caution and adversarial relationships (Thompson *et al.*, 1998; Moore and Dainty, 2001; Dubois and Gadde, 2002; Koskela, 2003). Industry reports (e.g. Egan, 1998) decried the prejudice-laden and adversarial nature of construction contracting as hampering innovation and performance and made sweeping recommendations, including the need to learn from other industries, especially manufacturing. There has been some scepticism about the specific thrusts of recent calls for change (e.g. Green, 1999; Male, 2003), and also about the comparability of construction with manufacturing and, hence, the potential for inter-industry learning (e.g. Green *et al.*, 2005; Winch, 2003).

Still, a second school of thought considers these recent initiatives as reflecting a shift of emphasis and purpose from a hitherto policy-driven construction industry to a market-driven industry, fuelled by the wave of privatisations that have swept across the world. For example, between the mid 1970s and the late 1980s, the public sector's share of total construction expenditure in the UK fell from over 50% to just over 20% (Murray and Langford, 2003). Not willing to accept construction as the loss-making part of their corporate endeavours, the majority of large private clients demanded and, largely, obtained strategic fit with their core business processes. Convinced about the 'value' prospects of these new ways of working, the public sector seems to have joined the bandwagon (NAO, 2001). Whatever the motivations for these reviews might be, it appears that the commercial realities of construction contracting today, that is, the need to meet the growing expectations of clients efficiently and effectively, make the reform agenda both necessary and largely irreversible. Cooperative strategies like partnering and alliancing have thus become paradigmatic (Pryke and Smyth, 2006).

However, the basis for predicting a relationship between cooperation and project performance has remained typically logical and conceptual rather than empirical. Cooperation lacks conceptual–definitional clarity and the discourse also tends to confound different levels of analysis (see Bresnen and Marshall, 2000a, 2000b; Fisher and Green, 2001). There is a dearth of research (Phua, 2004) addressing the socio-psychological factors that determine an individual's cooperative behaviour in construction and none, to the best of the authors' knowledge, which has considered economic incentives and socio-psychological factors as variables in the same study. These observations reflect a wider practice where industry solutions to major problems are usually of an *ad hoc*, piecemeal nature involving a number of elements to be 'bolted' on to already ailing systems (Kumaraswamy, 1998), with academia mostly playing catch up (Tookey *et al.*, 2001). While these are teething problems in its development as a field of enquiry (Parsons and Shils, 1959), construction management research needs to move on from *ad hoc* classification systems and taxonomies to comprehensive conceptual frameworks and theory-based systems.

This chapter outlines a research project that uses a theory-based approach to investigate the factors that shape individuals' cooperative behaviours in construction project scenarios. The discussion is structured into four sections. The first section discusses definitions for, and differentiates between, cooperation and collaboration. The following section outlines the theoretical framework that guides the discussion of this chapter. Next, the authors review current research on cooperation in construction management research and argue the need for an integrated framework. A set of remarks and recommendations concludes the chapter.

Defining cooperation

Cooperation as a concept is usually subsumed under notions about coordination and integration, which arise from interdependence between organisations or their members (Lawrence and Lorsch, 1967; Thompson, 1967; Van de Ven *et al.*, 1976; Argote, 1982; Tjosvold, 1984). Cooperation is a loosely defined term in the construction management literature and is often confused with collaboration. Thompson and Sanders (1998) used the terms *cooperation* and *collaboration* to describe different degrees of alignment/integration in the partnering relationship between organisations. They conceptualised cooperation and collaboration as opposites to competition, with collaboration signifying a higher degree of integration than cooperation. Love *et al.* (2002), citing other researchers (e.g. Hamel *et al.*, 1989; Kanter, 1994), used the terms cooperation and collaboration to distinguish the longevity of alliances between organisations. They considered long-term alliances as *cooperative* and short-term alliances as *collaborative*. The majority of project-related researchers seem to use cooperation and collaboration interchangeably (e.g. Bresnen and Marshall, 1998; Vaaland, 2004). Phua and Rowlinson (2004:45) conclude that 'cooperation' has been 'transformed into the much touted partnering relationship'. This variability in terminology makes it difficult to relate and integrate the different contributions of scholars researching projects. The *Oxford Dictionary of English* (Oxford University Press, 2005) defines cooperation as *the action or process of working together to the same end* and collaboration as *the action of working with someone to produce something*. From these, it is clear that *collaboration* is distinct from *cooperation*. The same inference is reached from a review of various definitions by leading researchers presented in Table 5.1 (Johnson, 1975; Schermerhorn, 1975; Pinto and Pinto, 1990; Ring and Van de Ven, 1994; Smith *et al.*, 1995; Wagner, 1995; Tyler and Blader, 2000). To illustrate, every construction project typically involves *collaboration* between clients, designers, constructors and facilities managers working together to deliver a construction product or service and this collaboration tends to be across, rather than within, organisational boundaries.

Arguably, the confusion in terminology is a by-product of efforts to 'learn' from other industries. Collaboration has been the construction industry's response to uncertainty and complexity (see Eccles, 1981; Winch 1989; Shirazi *et al.*, 1996). In manufacturing, however, collaboration is a relatively new concept and firms use collaboration typically to gain competitive advantage (see Hamel *et al.*, 1989; Kanter, 1994; Green *et al.*, 2005). The motivations are thus different: in manufacturing *collaboration* is used for competitive positioning while in construction it is required to integrate the contributions of differentiated functions. Recognition of the collaborative nature of construction contracting has been elusive. Dulaimi *et al.* (2003:309) argue that each collaborating partner in a construction project has every right to 'pursue their own interests, sometimes even at the expense of others'. The character of collaboration in construction and manufacturing is similar. In manufacturing, collaboration is competition in a different form – a competitive strategy for firms – and can lead to competitive compromise (Hamel *et al.*, 1989). In construction, the collaborating firms are not necessarily in direct competition with one another since they would need to be selected on the basis of complementarity. However, there is considerable scope for them to have separate objectives, the pursuit of which sets them in competition with the project's objectives. In both contexts, cooperation develops from a similarity or congruence of objectives and competition from a divergence of

Table 5.1 Definitions of cooperation

Author(s)	Term	Definition	Type(s)/dimension(s)
Schermerhorn (1975: 47)	Inter-organisational cooperation	The presence of deliberate relations between otherwise autonomous organisations for the **joint accomplishment of individual operating goals**	
Johnson (1975:241)	Cooperation	The coordination of behaviours among individuals **to achieve mutual goals**	
Ring and Van de Ven (1994:96)	Cooperative inter-organisational relationships	Socially contrived mechanisms **for collective action**, which are continually shaped and restructured by actions and symbolic interpretations of the parties involved	Definition based on a developmental process perspective
Smith, Carroll and Ashford (1995:10)	Cooperation	The process by which individuals, groups and organisations come together to interact and form psychological relationships **for mutual gain or benefit**	*Formal* – derive from contractual obligations and formal structures of control *Informal* – adaptable arrangements in which behavioural norms determine contribution of the parties
Tyler and Blader (2000:3)	Cooperation	Refers to whether or not people act **to promote the goals of the group**	*Mandatory* – required by group rules, job description *Discretionary* – not required by organisational norms or rules
Wagner (1995:152)	Cooperation	The wilful contribution of personal effort to **the completion of interdependent jobs**	*Voluntary* cooperation
Pinto and Pinto (1990:203)	Cross-functional cooperation	The quality of task and interpersonal relations when different functional areas work **to accomplish organisational tasks together**	

objectives (Pruitt and Carnevale, 1993). Therefore, *cooperation* refers to a specific quality of this collaboration – that of goal congruence (Tjosvold, 1984; Kanter, 1994; Bennett and Jayes, 1995; Dulaimi *et al.*, 2003). This concept of cooperation thus refers to *behaviour* that benefits all parties.

Cooperation is also a *process* and from that perspective is dynamic (Ring and Van de Ven, 1994; Smith *et al.*, 1995). This highlights the importance of time and the potential for organisational learning (see Sherif *et al.*, 1954). Based on this, the following is offered as a *possible* explanation for Love *et al.*'s (2002) classification of strategic alliances in construction: a long-term alliance offers greater scope for objective alignment through inter-organisational learning than a short-term alliance, which could be just 'business as usual' (i.e. collaboration) (see Hamel *et al.*, 1989; Kanter, 1994). Thus far, the analysis has been at the organisational level and cooperation has conceptually been shown to be opposed to competition. The analysis is, however, different at the individual level. Cooperation between different companies ultimately reduces to cooperation between individual managers from these companies (Tjosvold, 1984; Bresnen, 1991; Smith *et al.*, 1995; Kamann *et al.*, 2006). Individuals are not in competition with the (project) organisation nor its authority figures for that matter (Tjosvold, 1984; Tyler and Blader, 2000). In this sense, the opposite of cooperation is not competition, but rather a lack of cooperation. At the individual level, cooperation is therefore a unidirectional construct. The interest is thus the extent to which individuals will cooperate with their organisations and the factors that shape these cooperative behaviours.

This micro-level analysis also requires studying individuals' behaviours in the work settings that have the proximate role in shaping their experiences. This discussion therefore focuses on individuals' cooperation with their proximate project work groups. The proximate work group is the primary medium through which shared climates emerge, evolve and become embedded into the fabric of the (project) organisation (see Ring and Van de Ven, 1994; Anderson and West, 1998; Tyler and Blader, 2000). On a construction project, an individual's proximate work group comprises colleagues from the different functions and disciplines with whom they interact regularly in order to perform project tasks. The definition proposed by Tyler and Blader (2000; see Table 5.1) is consistent with the foregoing discussion and fits the purposes of this chapter. Cooperation is therefore defined as *behaviour that promotes the goals of the work group to which one belongs*. This definition recognises that while individuals' long-term interests are aligned with work group (and, hence, project) goals, their short-term self-interests may conflict with work group goals. The pursuit of individuals' long-term self-interests thus requires the subordination of short-term self-interests (Komorita and Parks, 1994; Tyler and Blader, 2000).

Cooperation in contracting

The above definition of cooperation requires conceptualisation of the *project organisation*. In Chapter 4, Kumaraswamy, Anvuur and Mahesh argue that that price, authority and trust are independent forms of control, which can be combined in various forms in setting up contractual frameworks or governance structures for construction projects. This still leaves open the question as to the objective for *contracting*. Turner and Müller (2003) argue that projects are a means of implementing change. In construction, contracting enables the client to form a temporary project organisation to implement this change. Resources are assigned to the project organisation to undertake the processes required and to manage uncertainty, complexity and need for integration in order to achieve the specified objectives. The *raison d'être* for contracting is, therefore, to create a cooperative project organisation to achieve the objectives of the change being contemplated (Turner and Simister, 2001). This is more so because a cooperative, rather than just a collaborative, project organisation minimises transaction costs under conditions of high uncertainty and complexity. Thus, contractual frameworks should be selected with an aim to align the motives of the contracting parties. This is the fundamental principle of incentivisation – 'shared risks unify motives' (see Bower *et al.*, 2002). Such a view adequately explains the empirical anomalies associated with attempts to apply transaction cost economics in construction (see Winch, 1989; Turner and Simister, 2001).

Contracting is thus conceptualised in Figure 5.1. This framework is adapted from Van der Geest (2004), whose analysis of livelihood strategies of subsistence farmers in Ghana mirrors construction contracting. His idea of *vulnerability* is akin to a construction industry characterised by uncertainty, complexity, hostility and poor performance (Gidado, 1996; Shirazi *et al.*, 1996; Dubois and Gadde, 2002). Vulnerability is determined by the degree of risk exposure, coping capacity and recovery potential (Van der Geest, 2004). Some of these risks are idiosyncratic to the particular projects considered while others are of a covariate nature, affecting all firms and projects in the industry. This collective vulnerability is determined by the institutional and market structures that limit the ability of firms to adequately deal with uncertainty and complexity. Coping capacity, recovery potential and, consequently, performance are determined by the project's 'capitals', which include:

- *Physical capital* – including raw materials, equipment and financial resources.
- *Human capital* – skills, expertise, professional competence and experience.
- *Structural capital* – procurement and contractual strategies, decision support procedures and systems, delivery modalities like partnering

and alliancing, 'professional and trilateral governance'. (Winch, 2001: 802–803)

- *Social capital* – 'communities of practice' (Shirazi *et al.*, 1996; Wenger, 1998; Wenger *et al.*, 2002), industry culture, social networks, experiences of the past, and expectations of the future.

These are all, or rather ought to be, within the purview of 'the management of projects' (Morris, 1994; Pryke and Smyth, 2006) and, acting individually and in concert, they shape (that is, enable, facilitate, enhance or constrain) a project's coping capacity, recovery potential and eventual performance. Differences in 'capitals' would therefore be reflected in the projects' coping capacity, recovery potential and performance.

The model in Figure 5.1 recognises that all projects have to deal with uncertainty and all but the simplest have to deal with complexity. Some uncertainty arises from the complexity of projects and some from seasonal variability (Eccles, 1981; Winch, 1989). Uncertainty that Winch (1989) calls *contracting uncertainty* can come from the governance systems. Firstly, there is considerable scope for discrepancy between the

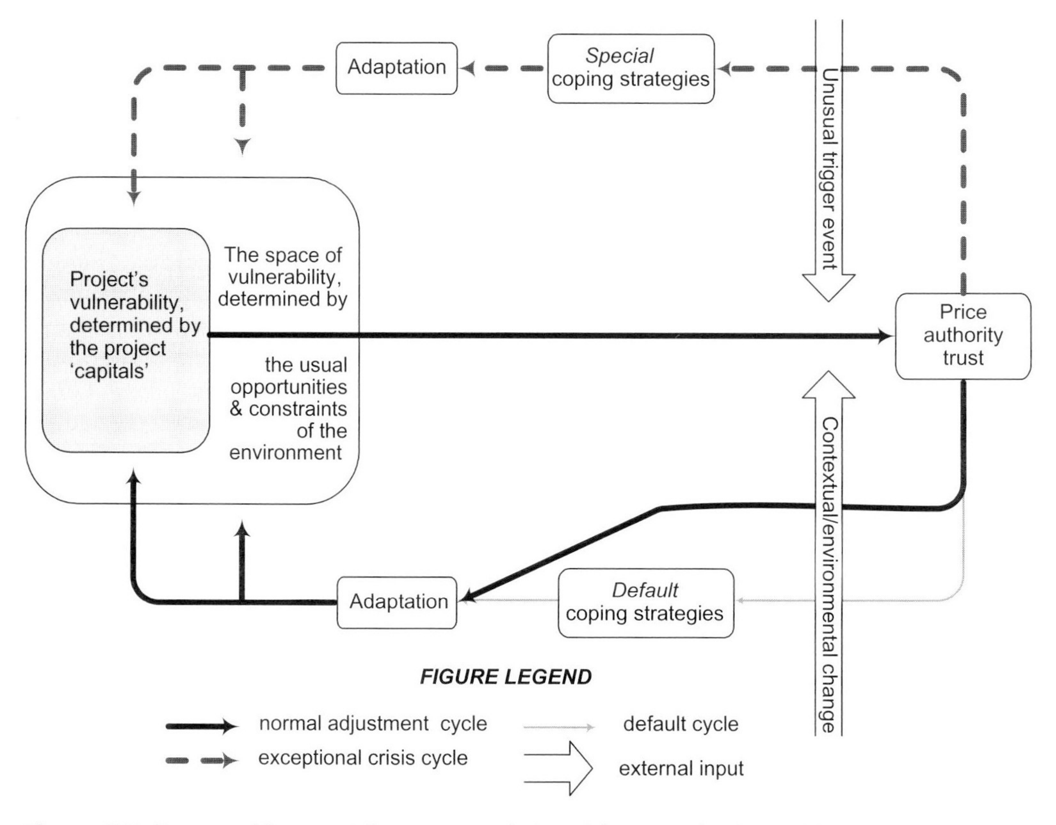

Figure 5.1 Conceptual framework for contracting (adapted from Van der Geest, 2004)

estimated and out-turn costs. Secondly, the price is itself in part dependent on the governance structures (Winch, 1989; Turner and Simister, 2001). Elaborate price adjustment mechanisms are thus a standard feature of construction contracts. The hierarchical nature of these price adjustment mechanisms makes them no different from coordination by adjustment of quantities (see Stinchcombe, 1985; Simon, 1991). Incentives and sanctioning mechanisms – important features of authority systems – are complicated by measurement problems arising from complexity and, thus, create further uncertainty; as does delivering a project to often incomplete performance specifications. The process of trust development creates uncertainties of its own (Oberschall and Leifer, 1986). These issues are usually addressed by adopting *default* coping strategies. These strategies include subcontracting, special procurement systems, 'professional governance', 'trilateral governance', 'communities of practice' and delivery modalities. Many of these management strategies have had very limited impact (see Koskela, 2003). It is suggested that this is because they are assembled and optimised solely upon (measurable) efficiency rather than broader notions of efficiency and effectiveness.

The model also takes into account periods of unusual stress when 'special' coping strategies are adopted. Such strategies are triggered by unusual events such as the Asian financial crisis of 1997 and the 'stop-go' policies of the British government towards construction in the 1970s. Typical responses to such unusual shocks take the form of zero-sum competitive tendencies and multi-layered subcontracting, which are counterproductive to the objective of creating and sustaining a cooperative project organisation (see Hillebrandt, 1984; Ball, 1988). In the absence of these sudden shocks, projects still have to cope with gradual changes in the opportunities and constraints of the environment. In addition to project uncertainty and complexity, this may involve dealing with problems arising from the casualisation of construction labour, gradual changes in construction technology and market (see Winch, 1989; Green *et al.*, 2005), or, indeed, responding to initiatives designed to reverse the tide of climate change. Adaptation thus involves both structural changes as a response to the nature and context of construction projects and coping with short-term shifts in the demand for construction. Different firms will have different coping capacities to the same shocks. For example, Green *et al.* (2005) conclude that the current stability of the UK construction market explains the recent interest in integrated supply chain management. Of course, if the effect of a sudden shock persists or is particularly adverse, some of the response strategies may become permanent features of the industry. This may explain, for example, the pervasive use of multi-layered sub-contracting in the Hong Kong construction industry.

Any study of construction project cooperation must therefore recognise the institutional and processual context of construction contracting.

This conceptualisation of contracting brings clarity to, and broadens the scope of, project management and, hence, the roles of project managers (see Turner and Müller, 2003). It also recasts hitherto competing research theories, such as the resource-based view (see Bridge and Tisdell, 2004, 2006; Chang, 2006), transaction cost economics approach (see Winch, 1989, 2001; Cox and Thompson, 1997; Thompson *et al.*, 1998; Walker and Chau, 1999), transformation, flow and value generation framework (see Koskela, 2003; Koskela and Ballard, 2006; Winch, 2006) and relationship marketing approach (see Davis, 1999; Low and Tan, 2002; Davis and Walker, 2004; Edkins and Smyth, 2006), as complementary in achieving the objectives of construction contracting.

Motivations for cooperative behaviour

This conceptualisation of contracting does not mean that transaction cost minimisation (or efficiency) is not an objective of construction projects. It just means that the efficiency criterion is not an adequate explanation for governance structures in construction. It is certainly still useful to know which transactions are 'costly' *vis-à-vis* the 'benefits' they confer. This view of project organisation allows us to address the concerns of flesh-and-blood human beings rather than treating them as robots (see Chapter 4 by Kumaraswamy, Anvuur and Mahesh). Cooperation emerges, evolves, grows and dissolves over time as a consequence of individual interactions between managers from the respective contracting parties (Ring and Van de Ven, 1994). Therefore, efficiency is only one outcome expectation. The second objective and outcome expectation is organisational justice: equity, procedural and interactional. Equity, as a principle of distributive justice, emphasises proportionality between investments made and benefits received. Procedural fairness relates to an individual's assessment of the fairness of decision-making procedures, and interactional justice, to the enactment of those procedures by the legitimate authorities. Procedural and interactional justice, or 'fair dealing' as they are otherwise referred to, are especially important because they shape people's judgements of the equitability and efficiency of their outcomes. These judgements influence people's motivations to cooperate (Ring and Van de Ven, 1994; Tyler and Blader, 2000).

The hierarchical nature of construction contracts allows for continuous mutual adjustment. This is important for maintaining equity and procedural fairness. The presence of 'professional governance' and 'trilateral governance' (Winch, 2001) improves perceptions of the legitimacy of authority figures. Parties are then more likely to judge decision-making processes and their outcomes as fair and consequently cooperate with the project organisation (Tyler and Blader, 2000). This cooperation has a stronger impact on efficiency than reducing transaction costs (Turner and Simister, 2001). However, things can sometimes go wrong and clients usually need some sort of restitution. This is typically achieved by incorporating some sanctioning mechanisms (e.g. retention funds, performance bonds and liquidated damages) and setting up surveillance regimes. These measures encourage strict contractual compliance and shift the focus from 'working the project' to 'working the contract' (Thompson *et al.*, 1998). But if focus is placed on encouraging cooperation, then these mechanisms become redundant as parties become self-regulating and motivated to advance the project's goals. It is also common for contracts to include clear systems of reward (i.e. performance incentives) tied to the achievement of specified performance targets.

Yet, the greatest advantage that a temporary organisation conceptualisation of contracting confers is, perhaps, the potential for *organisational identification*. Organisational identification is the cognitive meshing of self and organisation (Mael and Ashforth, 1992; Tyler and Blader, 2000). This cognitive connection to the organisation encourages behavioural engagement, that is, cooperation. This type of cooperation has been shown to result in a focus on quality, creativity and innovation (e.g. Mahaney and Lederer, 2006). For a comprehensive review of the creativity literature, the reader is referred to Crant (2000) and Shalley *et al.* (2004). Organisational identification is probably the most important reason why classical economics cannot explain the firm (see Simon, 1991). Identification is a potent force for resolving the externalities produced by attachment to sub-goals, by virtue of the commitment it can generate to project goals. The extensive lines of literature on intergroup relations identify equity, procedural and interactional justice as primary antecedents of organisational identification (see Gaertner, *et al.* 1993; Pettigrew, 1998). People with high organisational identification tend to evaluate their leaders more in terms of their procedural and interactional justice and less in terms of the equitability or favourability of their decisions and policies (Tyler and Blader, 2000).

A temporary organisation view of construction contracting offers great insight as to why many management strategies have failed or achieved only marginal success. Well intentioned integrative procurement arrangements have had very limited impact because they ignore process. The assessment and recommendation of alternative procurement arrangements are based on output variables, especially time, cost and quality. Whatever impacts the structural features of these procurement arrangements may have on performance, the hypothesised outcomes are realisable only to the extent that there is sufficient cooperation between the contracting parties (see Pocock *et al.*, 1996, 1997; Ngowi, 2000; Koskela, 2003; Cicmil and Marshall, 2005). Therefore, it is not surprising that clients pay lip service to selection methodologies when choosing procurement options and are willing, and are sometimes even eager, to modify or amend these to meet their needs (Bowen *et al.*, 1997; Tookey *et al.*, 2001). Contractual systems selected solely on the basis of cost, schedule and quality performance frequently create conflicts of motives between the contracting parties and can be counterproductive (see Winch, 1989; Bower *et al.*, 2002).

This conceptualisation also helps to explain emerging trends in contracting practices of procuring agents. Partnering, alliancing, supply chain management and concurrent engineering are some of the renewal strategies proposed to solve the adversarial nature and poor performance of construction contracting. While they have structural features that distinguish them from one another, they all use the team approach (cf. Katzenbach and Smith, 1993) as the primary means to execute their respective philosophies. As argued in detail elsewhere (Anvuur and Kumaraswamy, in press), organisational identification is the ultimate aim of these strategies. This does not require project actors to forsake their original identities (i.e. identification with their respective firms). Instead, it entails recategorising the memberships to create a superordinate identification with the temporary project organisation (Gaertner *et al.* 1993). We suggest that this effort can be described in terms of the widely used construct of project culture. In addition to creating taxonomies, this effort is, more importantly, concerned with understanding the aspects of intergroup structures and processes that create and sustain the attitudes and values that encourage cooperation on construction projects.

A theory-based conceptualisation of cooperation

This sub-section links the foregoing discussion to motivational theories and then outlines the dimensions of the cooperation construct as the bases for a metric. These dimensions of the cooperation construct have the added benefit of having already been validated in previous research. It is clear from the discussions above that there are two types of motivation for cooperation: internal and extrinsic motivations. This is based on Lewin and Gold's (1999) model of motivation for behaviour (i.e. *behaviour* = *f*[person, environment]). The extrinsic factors stem from contractual obligations and formal structures of control (e.g. job design and description). A proper definition of job roles ensures work-group members' efforts are channelled into undertaking only what is necessary to achieve the organisation's goals. By undertaking the behaviours mandated by their job roles, members are cooperating with their organisations. Such cooperation is thus called obligatory cooperation (Smith *et al.*, 1995; Tyler and Blader, 2000). The idea that job performance behaviours can be cooperative may not appear that intuitive. However, role definitions are rarely exhaustive. On construction projects, complexity, and the flexibility required, make exhaustive role definition not only impossible, but also undesirable. Individual managers therefore have considerable latitude in how they execute their project roles. They can carry out their work with a focus on doing just the minimum required to get by. Alternatively, they can carry out the same functions with zeal, vigour and a focus on quality. When managers contribute much more in job performance than the minimum, they are cooperating with the project organisation (Simon, 1991; Tyler and Blader, 2000).

These job roles are usually linked to performance incentives and sanctioning mechanisms. These create expectations of reward for a job well done and punishment for rule-breaking (Locke and Latham, 1990). The underlying assumptions of performance incentive and sanctioning systems emanate from social exchange theory. According to this theory, people interact in order to exchange resources and will seek to maximise personal gain in such interactions (Lewin and Gold, 1999). Therefore, individuals might cooperate to the extent that their outcomes, in terms of rewards and sanctions, are influenced by their cooperation. However, the performance incentives and sanctioning mechanisms in contracts operate at the level of the contracting firms and not the individual managers representing them. The motivational force they will exert on managers is thus predicated on the strength of their identification with the goals of their respective firms.

The second aspect of Lewin and Gold's (1999) model is internal motivation. Phua (2004:1034) defines this as the 'intrinsic and voluntary leanings of individuals to wilfully contribute their personal efforts to the completion of interdependent jobs'. These motivations flow from people's attitudes and values. Attitudes are the affective, cognitive and behavioural reactions towards the workgroup. Intrinsic motivation (Deci, 1975) and organisational identification (Hogg and Abrams, 1988) are two important attitudes that lead individuals to want to engage in actions that promote the welfare of their workgroups. Values reflect a feeling of responsibility to follow workgroup rules. Individuals internalise values and freely conform to workgroup norms. These internal motivations are explained by social identity theory (Tajfel, 1978, 1982; Tajfel and Turner, 1979). According to this theory, all individuals need to project a positive sense of self-identity. Groups help people to create and maintain a sense of self and a feeling of self-worth. According to Tajfel and Turner (1979), this need for positive self-identity motivates social comparisons that favourably differentiate ingroup from outgroup. Ingroup members are viewed as being very similar to self, are evaluated more positively and receive greater pro-social (i.e. helping) behaviour than members of the outgroup.

On construction projects, organisational identification takes the form of superordinate identification with the proximate work group. As discussed above, the saliency of this common ingroup identity is shaped by perceptions of justice. The behaviours flowing from these internal motivations are, therefore, voluntary in nature (Smith *et al.*, 1995; Tyler and Blader, 2000). They are voluntary because they come from within

Table 5.2 Types of cooperative behaviour (adapted from Tyler and Blader, 2000)

Functions of the behaviour	Forms of cooperative behaviour	
	Obligatory	Voluntary
Advancing the (project) organisation's goals	In-role	Extra-role
Restraining behaviours that harm the (project) organisation	Compliance	Deference

the individual, not the result of environmental contingencies. This view assumes instrumental motives to be primarily economic in nature. However, it is not uncommon for researchers to extend the resource exchange perspective to include non-material resources like self-satisfaction and self-esteem. The danger is that this tends towards a theory of instrumentality that is so broad and, therefore, scientifically unproductive (Montada, 1996; Tyler and Blader, 2000).

The foregoing discussion leads to a four-component theory-based measure of cooperation shown in Table 5.2. First, there are cooperative behaviours that directly advance the project's goals. When mandated by job roles, these are job performance behaviours or *in-role* behaviours (O'Reilly and Chatman, 1986). Cooperative behaviours that are not directly specified by a job description but which contribute directly to achieving project goals are contextual performance or 'organisational citizenship' behaviours. In other words, they are *extra-role* behaviours (O'Reilly and Chatman, 1986; Organ, 1988). Empirical evidence that in-role and extra-role behaviours are conceptually different and have independent influences on organisational performance is provided in many studies (e.g. Motowidlo and Van Scotter, 1994; Orr *et al.*, 1989).

Then there are behaviours that contribute indirectly to the accomplishment of project goals. This function of cooperation is concerned with discouraging behaviour that is harmful to the project organisation or inhibits its smooth functioning. Such cooperative behaviour is, therefore, restraining behaviour (Tyler and Blader, 2000). These behaviours are usually defined in the context of project work rules, which also spell out applicable sanctions for any breach of these rules. When members obey work rules because not doing so may lead to their detection and punishment, they are complying with these rules since they construe them as being obligatory. Such cooperative behaviour is called *compliance*. However, people can also submit to these rules because they feel it is the right and proper thing to do. Such voluntary rule following is called *deference*. Tyler and Blader (2000) demonstrate that compliance and deference are conceptually distinct dimensions of rule-following behaviour.

Cooperation research in construction

Managers can use the two major types of motivation to gain cooperation on construction projects. To the extent that they exercise control over resources and/or instruments of surveillance and sanctioning, project managers can influence people's behaviours. Project managers can also tap into people's internal motivations by appealing to or creating the attitudes and values that underpin people's behavioural intentions. Research shows that people's internal motivations can be shaped by influencing the context (i.e. job characteristics, work setting, relationships with peers and supervisors) in which people work (e.g. Tyler, 2002; Shalley *et al.*, 2004). In construction management research, this point is underscored in discourses on 'project culture' and 'project chemistry'. Success in this regard requires proactive project leadership, which is an important aspect of 'the management of projects' (Morris, 1994; Fellows *et al.*, 2003; Liu *et al.*, 2003; Turner and Müller, 2003; Liu and Fang, 2006) that has received very limited attention (Smyth and Edkins, 2007).

Situational factors influence the development and use of extrinsic and internal motivation on construction projects. For instance, performance incentive and sanctioning systems are difficult to administer in environments characterised by uncertainty and complexity (Ashley and Workman, 1986; Simon, 1991; Kohn, 1993; Thompson *et al.*, 1998; Fernie *et al.*, 2006). Also, clear systems of reward undermine individuals' intrinsic motivation (Deci, 1975). Therefore trade-offs exist in the use of these two types of motivation. The relative predominant magnitudes of cooperation from the two sources should weigh heavily in any trade-off. Research in (other) work organisations suggests that most of the variance in people's cooperative behaviour is explained by their internal motivations and *not* the desire to win rewards or avoid punishment (e.g. Tyler and Blader, 2000). The discussions in this chapter shed some light on factors that are important in creating the 'right' project context that appeals to or facilitates the development of people's attitudes and values that underpin their cooperative behaviours. But these need to be investigated directly in construction management research. While some support for the independent effects of the two motivations for cooperation exists in the context of the construction industry (Ashley and Workman, 1986; Phua, 2004; Kadefors, 2005), construction management research has yet to explore performance incentive/sanctioning mechanisms with socio-psychological factors as variables in the same study.

The importance of cooperative relationships to construction project success is highlighted in many studies. However, evidence of the link between cooperation and project performance has remained largely intuitive and logical, rather than empirical. Few studies have directly measured cooperation, most preferring to rely on project performance outcomes (especially time, cost and quality) as surrogates or indicators of cooperation on projects. Such feedback only leads to reactive project management (see also Chapter 6 by Smyth). Also these project performance outcomes confound the effects and influences of many factors other than cooperation. Two studies are acknowledged as relevant. Using a grounded empirical approach, Phua and Rowlinson (2004) demonstrated that the cooperation construct in their study was the most important determinant of project success, explaining 28% of the variance. However, their measure of cooperation considered only one dimension (*extra-role* behaviours) of the construct. Also, many other factors identified in their study were antecedents or correlates of cooperation and therefore present a possible problem of co-linearity.

Many studies investigating the effect of procurement arrangements on project performance outcomes report very weak or an absence of relationships. In a study of 234 (i.e. 25 and then later, 209) military construction projects, Pocock *et al.* (1996, 1997) investigated the relationship between 'degree of interaction' (DOI), a proxy for integration, and project performance outcomes. They found that projects with high levels of (cooperative) interaction had better and consistent performance than those with low DOIs. Pocock *et al.*'s (1997) findings suggest that procurement approach may have an indirect effect on project performance through cooperation. More specifically, procurement options provide differing opportunities for cooperative interaction, which is essential for the workgroup learning processes that lead to superordinate identification (Gaertner *et al.*, 1993; Pettigrew, 1998). While yielding useful data, the above two studies and others (e.g. Smyth and Edkins, 2007) point to the usefulness of a theory-based and robust approach to the study and measurement of cooperation.

The many calls for cooperation strategies in construction reflect the industry's desire to move away from a past dogged by adversarial relationships and poor performance. However, research on the concept of cooperation has lagged behind such efforts. A robust research framework, as proposed, is required in order to study the theoretical underpinnings and effectiveness of these cooperation strategies. Perhaps even more important is the need to understand the reasons for individuals' cooperative behaviours on construction projects. This information will help project managers to promote initiatives that foster greater engagement in project workgroups. This will remove the 'hit or miss' element in many of these management initiatives. While cross-sectional studies are useful in providing statistics on the strengths of any superficial relationships between constructs investigated, longitudinal case studies are especially important to gain fundamental understanding of the structures and processes involved in construction contracting (see also Chapter 4 by Kumaraswamy, Anvuur and Mahesh).

Conclusions

This chapter discusses and resolves ambiguities in previous approaches and provides clarity in the conceptualisation of cooperation. Based on a conceptualisation of projects as temporary organisations, it outlines the theoretical domain of cooperation at the level of the individual managers and identifies the theory-based dimensions of the cooperation construct as the bases for a metric. There has been very little research specifically focused upon cooperation in construction projects. More general research on cooperation strategies is lacking construct validity. It is recommended that empirical research based on the integrative framework proposed in this chapter is undertaken to extend and complement existing general research on cooperation in construction projects. The findings of such research should help project managers understand the drivers and dynamics for members' engagement in project work groups as well as the structural contexts that facilitate this process.

References

Anderson, N.R. and West, M.A. (1998) Measuring climate for work group innovation: development and validation of the team climate inventory. *Journal of Organizational Behaviour*, 19(3), 235–258.

Anvuur, A.M. and Kumaraswamy, M.M. (in press) A conceptual model of partnering and alliancing. *Journal of Construction Engineering and Management.*

Argote, L. (1982) Input uncertainty and organizational coordination in hospital emergency units. *Administrative Science Quarterly*, 27(3), 420–434.

Ashley, D.B. and Workman, B.W. (1986) *Incentives in Construction Contracts.* A Report to the Construction Industry Institute (CII) Source Document 8, The University of Texas, Austin.

Ball, M. (1988) *Rebuilding Construction: Economic Change and the British Construction Industry.* Routledge, London.

Bennett, J. and Jayes, S. (1995) *Trusting the Team: the Best Practice Guide to Partnering in Construction.* Centre for Strategic Studies in Construction/Reading Construction Forum, University of Reading.

Bowen, P.A., Hindle, R.D. and Pearl, R.G. (1997) The effectiveness of building procurement systems in the attainment of client objectives. In: Davidson, C.H. and Meguid, T.A.A. (eds.) *International Symposium of Commission W92 CIB: Procurement – A Key to Innovation.* University of Montreal, Montreal, pp. 39–49.

Bower, D., Ashby, G., Gerald, K. and Smyk, W. (2002) Incentive mechanisms for project success. *Journal of Management in Engineering*, 18(1), 37–43.

Bresnen, M. (1991) Construction contracting in theory and practice: a case study. *Construction Management and Economics*, 9, 247–263.

Bresnen, M. and Marshall, N. (1998) Partnering strategies and organizational cultures in the construction industry. In: Hughes, W. (ed.) *14th Annual ARCOM Conference*, 9–11 September, Association of Researchers in Construction Management University of Reading, 2, 465–476.

Bresnen, M. and Marshall, N. (2000a) Motivation, commitment and the use of incentives in partnerships and alliances. *Construction Management and Economics*, 18(5), 587–598.

Bresnen, M. and Marshall, N. (2000b) Partnering in construction: a critical review of issues, problems and dilemmas. *Construction Management and Economics*, 18(2), 229–237.

Bridge, A.J. and Tisdell, C. (2004) The determinants of the vertical boundaries of the construction firm. *Construction Management and Economics*, 22(8), 807–825.

Bridge, A. and Tisdell, C. (2006) The determinants of the vertical boundaries of the construction firm: response. *Construction Management and Economics*, 24(3), 233–236.

Chang, C.-Y. (2006) The determinants of the vertical boundaries of the construction firm: comment. *Construction Management and Economics*, 24, 229–232.

Cicmil, S. and Marshall, D. (2005) Insights into collaboration at the project level: complexity, social interaction and procurement mechanisms. *Building Research and Information*, 33(6), 523–535.

Cox, A. and Thompson, I. (1997) 'Fit for purpose' contractual relations: determining a theoretical framework for construction projects. *European Journal of Purchasing and Supply Management*, 3(3), 127–135.

Crant, J.M. (2000) Proactive behaviour in organisations. *Journal of Management*, 26(3), 435–462.

Davis, P.R. (1999) Relationship marketing in the construction industry. *AACE International Transactions*, PM11, 1–6.

Davis, P.R. and Walker, D.H.T. (2004) Relationship based procurement. In: Khosrowshahi, F. (ed.) *20th Annual ARCOM Conference*, 1–3 September, Association of Researchers in Construction Management Heriot Watt University, 2, 887–895.

Deci, E.L. (1975) *Intrinsic Motivation*. Plenum Press, New York.

Dubois, A. and Gadde, L.-E. (2002) The construction industry as a loosely coupled system: implications for productivity and innovation. *Construction Management and Economics*, 20(7), 621–631.

Dulaimi, M.F., Ling, F.Y.Y. and Bajracharya, A. (2003) Organizational motivation and inter-organizational interaction in construction innovation in Singapore. *Construction Management and Economics*, 21(3), 307–318.

Eccles, R.G. (1981) Bureaucratic versus craft administration: the relationship of market structure to the construction firm. *Administrative Science Quarterly*, 26(3), 449–469.

Edkins, A.J. and Smyth, H.J. (2006) Contractual management in PPP projects: evaluation of legal versus relational contracting for service delivery. *Journal of Professional Issues in Engineering Education and Practice*, 132(1), 82.

Egan, J. (1998) *Rethinking Construction*. HMSO, London.

Fellows, R., Liu, A. and Fong, C.M. (2003) Leadership style and power relations in quantity surveying in Hong Kong. *Construction Management and Economics*, 21(8), 809–818.

Fernie, S., Leiringer, R. and Thorpe, T. (2006) Change in construction: a critical perspective. *Building Research and Information*, 34(2), 91–103.

Fisher, N. and Green, S. (2001) Partnering and the UK construction industry the first ten years – a review of the literature. In: NAO (ed.) *Modernising Construction*. National Audit Office, London, pp. 58–66.

Gaertner, S.L., Dovidio, J.F., Anastasio, P.A., Bachman, B.A. and Rust, M.C. (1993) The common ingroup identity model: recategorisation and the reduction of intergroup bias. *European Review of Social Psychology*, 4, 1–26.

Gidado, K.I. (1996) Project complexity: the focal point of construction production planning. *Construction Management and Economics*, 14(3), 213–225.

Green, S.D. (1999) Partnering: the propaganda of corporatism? *Journal of Construction Procurement*, 5(2), 177–186.

Green, S.D., Fernie, S. and Weller, S. (2005) Making sense of supply chain management: a comparative study of aerospace and construction. *Construction Management and Economics*, 23(6), 579–593.

Hamel, G., Doz, Y.L. and Prahalad, C. (1989) Collaborate with your competitors – and win. *Harvard Business Review*, January–February, 133–138.

Hillebrandt, P.M. (1984) *Analysis of the British Construction Industry*. Macmillan, London.

Hogg, M.A. and Abrams, D. (1988) *Social Identifications: a Social Psychology of Intergroup Relations and Group Processes*. Routledge, London.

Johnson, D.W. (1975) Cooperativeness and social perspective taking. *Journal of Personality and Social Psychology*, 31, 241–244.

Kadefors, A. (2005) Fairness in interorganisational project relations: norms and strategies. *Construction Management and Economics*, 23(8), 871–878.

Kamann, D.-J.F., Snijders, C., Tazelaar, F. and Welling, D.T. (2006) The ties that bind: buyer-supplier relations in the construction industry. *Journal of Purchasing and Supply Management*, 12(1), 28–38.

Kanter, R.M. (1994) Collaborative advantage: the art of alliances. *Harvard Business Review*, July–August, 96–106.

Katzenbach, J.R. and Smith, D.K. (1993) The discipline of teams. *Harvard Business Review*, March–April, 111–120.

Kohn, A. (1993) Why incentive plans cannot work. *Harvard Business Review*, September–October, 54–63.

Komorita, S.S. and Parks, C.D. (1994) *Social dilemmas*. Brown & Benchmark, Madison.

Koskela, L. (2003) Is structural change the primary solution to the problems of construction? *Building Research and Information*, 31(2), 85–96.

Koskela, L. and Ballard, G. (2006) Should project management be based on theories of economics or production? *Building Research and Information*, 34(2), 154–163.

Kumaraswamy, M.M. (1998) Industry development through creative project packaging and integrated management. *Engineering, Construction and Architectural Management*, 5(3), 228–237.

Lawrence, P.R. and Lorsch, J.W. (1967) *Organization and Environment: Managing Differentiation and Integration*. Division of Research, Graduate School of Business Administration, Harvard University, Boston.

Lewin, K. and Gold, M. (1999) *The Complete Social Scientist: a Kurt Lewin Reader*, 1st Edition. American Psychological Association, Washington, DC.

Liu, A. and Fang, Z. (2006) A power-based leadership approach to project management. *Construction Management and Economics*, 24(5), 497–507.
Liu, A., Fellows, R. and Fang, Z. (2003) The power paradigm of project leadership. *Construction Management and Economics*, 21(8), 819–829.
Locke, E.A. and Latham, G.P. (1990) *A Theory of Goal Setting & Task Performance.* Prentice Hall, Englewood Cliffs.
Love, P.E.D., Irani, Z., Cheng, E. and Li, H. (2002) A model for supporting inter-organizational relations in the supply chain. *Engineering, Construction and Architectural Management*, 9(1), 2–15.
Low, S.P. and Tan, S.L.G. (2002) Relationship marketing: a survey of QS firms in Singapore. *Construction Management and Economics*, 20(8), 707–721.
Mael, F. and Ashforth, B.E. (1992) Alumni and their alma mater: a partial test of the reformulated model of organizational identification. *Journal of Organizational Behaviour*, 13(2), 103–123.
Mahaney, R.C. and Lederer, A.L. (2006) The effect of intrinsic and extrinsic rewards for developers on information systems project success. *Project Management Journal*, 37(4), 42–54.
Male, S. (2003) Faster building for commerce: NEDO (1988) In: Murray, M. and Langford, D. (eds.) *Construction Reports 1944–98.* Blackwell Science, Oxford, pp. 130–144.
Montada, L. (1996) Trade-offs between justice and self-interest. In: Montada, L. and Lerner, M.J. (eds.) *Current Societal Concerns about Justice.* Plenum Press, New York, pp. 259–275.
Moore, D.R. and Dainty, A.R.J. (2001) Intra-team boundaries as inhibitors of performance improvement in UK design and build projects: a call for change. *Construction Management and Economics*, 19(6), 559–562.
Morris, P.W.G. (1994) *The Management of Projects.* Thomas Telford, London.
Motowidlo, S.J. and Van Scotter, J.R. (1994) Evidence that task performance should be distinguished from contextual performance. *Journal of Applied Psychology*, 79(4), 475–480.
Murray, M. and Langford, D.A. (eds.) (2003) *Construction Reports 1944–98.* Blackwell Science, Oxford.
NAO (2001) *Modernising Construction.* Report by the Controller and Auditor General: HC 87 Session 2000–2001, National Audit Office, London.
Ngowi, A.B. (2000) Construction procurement based on concurrent engineering principles. *Logistics Information Management*, 13(6), 361–368.
Oberschall, A. and Leifer, E.M. (1986) Efficiency and social institutions: uses and misuses of economic reasoning in sociology. *Annual Review of Sociology*, 12, 233–253.
O'Reilly, C.I. and Chatman, J. (1986) Organisational commitment and psychological attachment: the effects of compliance, identification, and internalization on prosocial behaviour. *Journal of Applied Psychology*, 71(3), 492–499.
Organ, D.W. (1988) *Organisational Citizenship Behaviour: the Good Soldier Syndrome.* Lexington Books, Lexington.
Orr, J.M., Sackett, P.R. and Mercer, M. (1989) The role of prescribed and non-prescribed behaviours in estimating the dollar value of performance. *Journal of Applied Psychology*, 74(1), 34–40.
Oxford University Press (2005) *The Oxford Dictionary of English*, Revised 2nd Edition. Oxford University Press, Oxford.

Parsons, T. and Shils, E.A. (1959) *Toward a General Theory of Action.* Harvard University Press, Cambridge.

Pettigrew, T.F. (1998) Intergroup contact theory. *Annual Review of Psychology*, 69, 65–85.

Phua, F.T.T. (2004) The antecedents of co-operative behaviour among project team members: an alternative perspective on an old issue. *Construction Management and Economics*, 22(10), 1033–1045.

Phua, F.T.T. and Rowlinson, S.M. (2004) How important is cooperation to construction project success? A grounded empirical quantification. *Engineering, Construction and Architectural Management*, 11(1), 45–54.

Pinto, M.B. and Pinto, J.K. (1990) Project team communication and cross-functional cooperation in new program development. *Journal of Product Innovation Management*, 7(3), 200–212.

Pocock, J.B., Hyun, C.T., Liu, L.Y. and Kim, M.K. (1996) Relationship between project interaction and performance indicators. *Journal of Construction Engineering and Management*, 122(2), 165–176.

Pocock, J.B., Liu, L.Y. and Kim, M.K. (1997) Impact of management approach on project interaction and performance. *Journal of Construction Engineering and Management*, 123(4), 411–418.

Pruitt, D.G. and Carnevale, P.J. (1993) *Negotiation in Social Conflict.* Open University Press, Buckingham.

Pryke, S.D. and Smyth, H.J. (eds.) (2006) *The Management of Complex Projects: a Relationship Approach.* Blackwell, Oxford.

Ring, P.S. and Van de Ven, A.H. (1994) Developmental processes of cooperative interorganizational relationships. *The Academy of Management Review*, 19(1), 90–118.

Schermerhorn, J.R., Jr. (1975) Determinants of interorganizational cooperation. *The Academy of Management Journal*, 18(4), 846–856.

Shalley, C.E., Zhou, J. and Oldham, G.R. (2004) The effects of personal and contextual characteristics on creativity: where should we go from here? *Journal of Management*, 30(6), 933–958.

Sherif, M., Harvey, O.J., White, B.J., Hood, W.R. and Sherif, C. (1954) *Experimental Study of Positive and Negative Intergroup Attitudes between Experimentally Produced Groups: Robbers Cave Experiment.* University of Oklahoma, Norman.

Shirazi, B., Langford, D.A. and Rowlinson, S.M. (1996) Organizational structures in the construction industry. *Construction Management and Economics*, 14(3), 199–212.

Simon, H.A. (1991) Organizations and markets. *The Journal of Economic Perspectives*, 5(2), 25–44.

Smith, K.G., Carroll, S.J. and Ashford, S.J. (1995) Intra- and interorganizational cooperation: toward a research agenda. *The Academy of Management Journal*, 38(1), 7–23.

Smyth, H.J. and Edkins, A.J. (2007) Relationship management in the management of PFI/PPP projects in the UK. *International Journal of Project Management*, 25(3), 232–240.

Stinchcombe, A. (1985) Contracts as hierarchical documents. In: Stinchcombe, A. and Heimer, C. (eds.) *Organisational Theory and Project Management.* Norwegian University Press, Bergen, pp. 121–171.

Tajfel, H. (1978) *Differentiation between Social Groups: Studies in the Social Psychology of Intergroup Relations.* Academic Press, London.

Tajfel, H. (1982) Social psychology of intergroup relations. *Annual Review of Psychology*, 33, 1–39.

Tajfel, H. and Turner, J.C. (1979) An integrative theory of intergroup conflict. In: Austin, W.G. and Worchel, S. (eds.) *The Social Psychology of Intergroup Relations.* Brooks/Cole, Monterey, pp. 33–47.

Thompson, I., Cox, A. and Anderson, L. (1998) Contracting strategies for the project environment. *European Journal of Purchasing and Supply Management*, 4(1), 31–41.

Thompson, J.D. (1967) *Organizations in Action: Social Science Bases of Administrative Theory.* McGraw-Hill, New York.

Thompson, P.J. and Sanders, S.R. (1998) Partnering Continuum. *Journal of Management in Engineering*, 14(5), 73–78.

Tjosvold, D. (1984) Cooperation theory and organizations. *Human Relations*, 37(9), 743–767.

Tookey, J.E., Hardcastle, C., Murray, M. and Langford, D. (2001) Construction procurement routes: re-defining the contours of construction procurement. *Engineering Construction and Architectural Management*, 8(1), 20.

Turner, J.R. and Simister, S.J. (2001) Project contract management and a theory of organisation. *International Journal of Project Management*, 19(8), 457–464.

Turner, J.R. and Müller, R. (2003) On the nature of the project as a temporary organization. *International Journal of Project Management*, 21(1), 1–8.

Tyler, T.R. (2002) Leadership and cooperation in groups. *The American Behavioural Scientist*, 45(5), 769–782.

Tyler, T.R. and Blader, S.L. (2000) *Cooperation in Groups: Procedural Justice, Social Identity and Behavioural Engagement.* Psychology Press, Philadelphia.

Vaaland, T.I. (2004) Improving project collaboration: start with the conflicts. *International Journal of Project Management*, 22(6), 447.

Van de Ven, A.H., Delbecq, A.L. and Koenig, R., Jr. (1976) Determinants of coordination modes within organizations. *American Sociological Review*, 41(2), 322–338.

Van der Geest, K. (2004) *'We're managing!' Climate Change and Livelihood Vulnerability in Northwest Ghana.* African Studies Centre, Leiden.

Wagner, J.A., III (1995) Studies of individualism-collectivism: effects on cooperation in groups. *The Academy of Management Journal*, 38(1), 152–172.

Walker, A. and Chau, K.W. (1999) The relationship between construction project management theory and transaction cost economics. *Engineering, Construction and Architectural Management*, 6(2), 166–176.

Wenger, E. (1998) *Communities of Practice: Learning, Meaning, and Identity. Learning in Doing: Social Cognitive and Computational Perspectives.* Cambridge University Press, Cambridge.

Wenger, E., McDermott, R.A. and Snyder, W. (2002) *Cultivating Communities of Practice: a Guide to Managing Knowledge.* Harvard Business School Press, Boston.

Winch, G.M. (1989) The construction firm and the construction project: a transaction cost approach. *Construction Management and Economics*, 7(4), 331–345.

Winch, G.M. (2001) Governing the project process: a conceptual framework. *Construction Management and Economics*, 19(8), 799–808.

Winch, G.M. (2003) How innovative is construction? Comparing aggregated data on construction innovation and other sectors – a case of apples and pears. *Construction Management and Economics*, 21(6), 651–654.

Winch, G.M. (2006) Towards a theory of construction as production by projects. *Building Research and Information*, 34(2), 164–174.

6 Developing trust

Hedley Smyth

Introduction

The need for increased levels of trust, flowing from a rallying cry for non-adversarial collaborative project working, was instigated under continuous improvement agendas, for example Egan (1998) in the UK. However, such rallying cries beg the question, 'What is trust?' Egan and many in his wake failed to say what is meant by trust and how it can be recognised. Anecdotally, attendance at London Cluster Meetings of M4i – an early vehicle for disseminating innovation – in the first years of the twenty-first century was witness to industry speaker after industry speaker emphasising the need for trust, saying, 'It's all about open communication.' One thing that trust is not all about is open communication. If there is complete transparency of communication then there is no need for trust. Collaborative relationships need trust, and complete transparency is simply unaffordable. Therefore, trust is needed in the face of uncertainty, hence a lack of information and information asymmetry. To develop collaborative relationships requires the development of trust. Developing trust in relationships requires an understanding of trust. This chapter aims to deepen this understanding by not only looking at definitions but also by showing how trust relates to:

- A dynamic set of concepts that provide the essential elements that are present for the formation and development of trust.
- A philosophical underpinning for trust formation, especially the relation of trust to moral philosophy and its operation in economic life.
- A methodological analysis that shows how trust is formed, including its contextual dependency on relationships, management and project management.

The chapter draws these aspects together within a framework of trust. An initial framework was first proposed for the understanding of trust (Smyth, 2003). However, critical analysis requires reflection and refinement, this chapter providing an opportunity to present how the framework has changed and, hence, improve the conceptual basis for understanding and applying to collaborative relationships in practice.

Therefore good intentions were behind the rallying cries for collaborative relationships based on trust, yet more is needed to develop and manage trust. Several factors influence the development of trust:

1. *Market change* – market structures and governance for *relational contracting* in alliances through partnering and supply chain management (e.g. Cox and Ireland, 2006; Kumaraswamy and Rahman, 2006), inducing behavioural reaction to the new conditions (Smyth, 2006a; Smyth and Pryke, 2006; cf. Pryke and Smyth, 2006), which have been the mainstay of most of the rallying cries for trust.
2. *Aggregate behaviour change* – behaviour and conduct arising from good intentions and ethical motives of active parties, which to date has been left to individual responsibility (Smyth, 2005; Edkins and Smyth, 2006).
3. *Proactive management of aggregate behaviour* – instigation of norms, systems, procedures and codes of behavioural conduct to conceptually develop *relationship management* (e.g. Grönroos, 2000; Gummesson, 2001; Ford *et al.*, 2003), which is currently marginal in contracting organisations and marginal within projects (Smyth and Edkins, 2007; cf. Pryke and Smyth, 2006).

Market change is insufficient to maintain long-term continuous improvement. *Proactive management* is necessary. The *aim* should be to develop trust conceptually, applying a framework of trust for developing trust in research and practice. The first framework proposed (Smyth, 2003) was used for measuring trust (Edkins and Smyth, 2006a; 2006b; Smyth, 2005, 2006a). Table 6.1 shows the initial elements and proposed elements in the revised framework of trust.

As proposed, the framework takes two dimensions into account:

1. *Philosophical* – linking the conceptual understanding of trust to a deeper philosophical understanding, derived from a broader literature review that covers a range of moral philosophical positions, namely virtue or character, intention and duty, and outcomes or performance in approaches of utilitarianism or nurture:
   i. Adding a new framework element, social obligations of trust, based upon intentions derived from duty and tolerance concerning dignity and respect.
   ii. Placing trust within the concept of the moral economy, which is foundational hence underpinning the market economy and is dispensational in contextually giving rise to ethical investment as social capital.
2. *Methodological* – relating trust to critical realism, which takes account of the causal drivers within the essence of the object (trust in relationships) and the contextual conditions mediating to produce a variety of general and particular outcomes or events:

Table 6.1 Framework for the understanding and analysis of trust

Initial elements	Proposed elements	Research references
Characteristics of trust	*Characteristics of trust*	Smyth, 2005, 2006a; Edkins and Smyth, 2006b
Components for trust	*Components for trust*	Smyth, 2006a; Edkins and Smyth, 2006a
–	*Social obligations of trust*	
Conditions of trust	*Conditions of trust*	Hannah, 1991; Thompson, 1996, 1998,; 2003; Smyth, 2005; Smyth and Thompson, 2005; cf. Smyth and Thompson, 1999
Levels of trust	*Levels of trust*	
Operational basis for trust	*Operational basis for trust*	Smyth, 2006b
Evidence of trust	–	
Trust in the marketplace	*Trust in the marketplace*	Smyth, 2005

i. Identifying four framework elements as largely ontological or essential for trust formation – characteristics of trust, components of trust, the proposed element social obligations of trust and conditions of trust – and assigning three elements as contextual conditions – levels of trust, operational basis for trust, and trust in the marketplace – omitting evidence of trust as it concerns events rather than conditions.
ii. Reflecting upon the paucity of causality in research on trust which tends to seek the general explanations under positivist and empiricist or particular explanations under empiricism and hermeneutics, rather than both in ways that account for contextual conditions, as offered by critical realism (see Appendix 1).

A specific definition of trust for business and projects was derived from an extensive review (Smyth, 2003):

> *Trust is a disposition and attitude concerning the willingness to rely upon the actions of or be vulnerable towards another party, under circumstances of contractual and social obligations, with the potential for collaboration.* (Smyth, 2005:212–213; Edkins and Smyth, 2006a:84)

A detailed exploration of the framework of trust elements is provided in the penultimate section, the following two sections expounding upon the philosophical and methodological issues that inform the framework.

Philosophical issues and trust

Trust was a poor relation in moral philosophy until recently (e.g. Baier, 1994; Gambetta, 1998). Historically, moral philosophy focused upon:

- Virtue or character: Aristotelian tradition – values, beliefs and attitudes formed from experience, contributing to individual and societal well-being, giving rise to a general disposition to trust according to the character of others and prevailing circumstances.
- Intentions and duty: Kantian tradition – informing actions to encourage individual and societal freedom, implying a willingness to trust and display trusting behaviour towards others.
- Outcomes or performance: utilitarian tradition *or* care tradition – *either* net benefits and welfare for individuals and society, where trust is largely incidental or relegated to subservient utility levels, *or*, individual bonding and social affirmation to reinforce individual identity, security and significance and improve performance and social contribution to society, leading to a position where trust is regarded as being an important part of nurture and care.

Traditions are linked (character forms values which informs intentions, influencing action, from which there are consequences either in goal-orientated utility or in process-orientated individual and social performance in business), yet there are also conflicts between them (Jones *et al.*, 2005), the significance of trust varying between traditions. Trust is both noun and verb, giving rise to interpretation difficulties between traditions. The noun trust concerns virtue, for example a belief (e.g. Shaw, 1997) or attitude (e.g. Luhmann, 1979; Flores and Solomon, 1998); the verb trust concerns disposition (e.g. Fukuyama, 1995) and behaviour (e.g. Moorman *et al.*, 1993; Currall and Judge, 1995; Mayer *et al.*, 1995; Smith and Barclay, 1995; Baier, 1994). Concerning a project, the noun trust refers to a snapshot of current conditions or circumstances, whereas the verb refers to trust development, maintenance, stasis or erosion.

Trust pertains to individuals in relationships. Does trust pertain to organisations, such as project teams and project-orientated businesses?

- People are different, thus trust (verb) in interpersonal relationships varies.
- Those individuals representing organisations will exhibit variance in trustworthiness (verb and noun) and, hence, cannot be reflecting organisational trust *per se*.
- Trust (noun) pertaining to organisational values does reflect trust to the extent these are embedded and are:
  - Expressed through strategy, including mission statements, 'core values', norms, bodies of knowledge, codes of conduct, quality

  - assurance policy, corporate social and sustainable responsibility measures.
  - And generally evident as goodwill, branding and reputation, and through product and service reliability.
- Trust towards an organisation (noun) is indirectly ascribed trust (verb), built up through aggregated behaviours, providing direct and indirect evidence for 'trusting' any organisation.

Therefore, organisational trust exists in different ways compared to interpersonal relationships. Trust has to be behaviourally embedded and/or actions that translate into moral values or into market value that are in evidence. So when people say they 'trust' an organisation, they are either directly trusting the aggregated behaviour experienced or indirectly trusting the goodwill, brand and reputation.

The trust of individuals has meaning within relationships, is affected by social obligations, and thus is essentially social.

Social obligations of trust

For the social obligations of trust to become a new framework element, two traditions of moral philosophy are considered: 'intention and duty' and 'nurture and care'. The social obligation of duty was a categorical imperative for Kant (1785). Duty, meaning individuals should consider others, is drawn from monastic asceticism and a virtue for its own sake (Norman, 1998), from which flow moral requirements to have dignity and respect for others as a matter of intention and duty. Dignity is embodied in the essence of being a person, acknowledging dignity of another is simply valuing them as a 'human being'. It is based in *equality* and cannot have a price attached. Dignity is enhanced through self-control overall and in prevailing circumstances. Such action leads to respect. Respect is earned, the value one person ascribes to another based upon past and current behaviour and achievement. Respect for another person may arise from direct experience or it may come indirectly through reputation. As respect is earned, it is therefore based in *equity* and associated with 'human doing', which can have a price attached to it in the job market.

The boundaries between 'intention and duty' and 'nurture and care' blur in practice. Occupying a position (trustee, board member, manager, project manager) carries authority and responsibilities with duties attached. Such positions carry status, which may lend dignity to the occupant, yet they have to earn respect to secure and maintain their position. Consequently, self-control can increase respect from others and enhance dignity. Conversely, actions causing loss of respect from others may erode dignity. The extent to which dignity helps to earn respect, respect enhances dignity or loss of respect erodes dignity, creates a social dynamic (which can filter into market price as a salary, or as the amount to be paid for a service) (Smyth, 2006b).

Gilligan (1982) introduced the idea of contextually based 'care' into moral philosophy. Her work, grounded in a feminist perspective, focused on bonding between mothers and their children as a nurturing process that brings development. Gilligan's enduring legacy to the field of moral philosophy has implications for management and project management. An important implication of the sense of nurture and care in organisations is the 'social capital' created by improved social and economic, inter-firm and intra-firm, relationships (Smyth and Pryke, 2006), which in turn is important in the context of developing collaborative working relationships in construction.

Whilst duty is important in management and relates to legal requirements, employment and supply contracts, culturally and through company rules, dignity and respect are not necessarily ascribed through dutiful requirement. Dignity and respect can be ascribed and acknowledged by choice. People tend to evaluate the character and behaviour of individuals, and respond accordingly. If parties are seen to care for others and nurture relationships, then they will generally receive respect and be treated with dignity beyond dignity and respect ascribed out of duty. Nurture therefore modifies the Kantian approach. Hence, social obligations regarding dignity and respect have become less concerned with duty and more concerned with love, caring support and tolerance. Whilst Kant was not legalistic in ascribing importance to duty, the approach of nurture adds more emphasis to grace and choice. It shifts the ground from being dependent upon intentions of being 'good' to choosing to be 'good' in the light of behavioural evidence – negotiated experientially. Yet the tradition of intention remains important, as one party must start the process by showing unconditional dignity towards another party.

A party will be more willing to trust another when they sense the other party values them as a person, showing dignity, which is reinforced when trust is reciprocated and both parties are allowed some dignity. Therefore, dignity and respect form a moral foundation for the creation of trust and responsibility for this comes through intention (inputs) and nurture (outputs). As such this forms a dynamic that contributes to the social obligations through societal culture and organisational culture. Recognition of the social obligations in any relationship is important for the potential and constraints of trust formation, hence the proposal for a new element, the *social obligations of trust*.

The moral economy and trust

Management philosophically is trying to encourage certain sorts of outcome. Management in terms of organisational behaviour is therefore trying to practically engender certain behaviours, including trust. This behavioural stance theoretically links to the relationship approach for managing projects (Pryke and Smyth, 2006; Smyth and Pryke, 2006).

Moreover, it takes collaborative project working from changing market structures and governance, *relational contracting*, towards proactive change in aggregate behaviour, developing *relationship management*. Trust is important in relational contracting and vital in relationship management, the kernel of both concepts being to potentially produce opportunities for positive behaviour. There is nothing intrinsically 'good' about trust (Baier, 1994) or in relationships (Gummesson, 2001) for they reinforce the dominant values and attitudes in the circumstances, depending upon what virtues, intentions, business and project goals are pursued. A fertile context is needed to create positive behaviour, which the concept of the *moral economy* provides (cf. Sayer, 2003).

It is suggested that morality is necessary for economic functioning. The *moral economy* runs in parallel, yet is linked to the market economy, having two elements:

1. *Foundational*, underpinning and facilitating the functioning of:
   i. The market economy as a whole.
   ii. Individual and serial transactions.
2. *Dispensational*:
   i. Being an era of a market economy.
   ii. Recognising moral behaviours are context specific in terms of management of the enterprise.
   iii. Managing the allocation or dispensing of resources and demonstrable actions to encourage certain moral behaviours as part of developing social capital and competencies of an organisation.

Foundationally the market economy requires some morality to operate, being enhanced in a dispensational sense by behaviour in the prevailing circumstances (Smyth, 2006b). Therefore the interests of business cannot simply be regarded as solely reliant upon and focused upon growth and profit. These are primary goals, yet with moral responsibility attached. Many managerial and economic models exclude morality or initially exclude and subsequently reintroduce morality in subsumed roles. For example, trust within transaction cost analysis and game theory works where the way forward is to 'bargain our way into morality' (Gaulthier, 1986, quoted in Rachels, 1995:148). Yet, markets are socially constructed (Polanyi, 1944) by skilled and specialist actors, who have social and moral responsibilities (Abolafia, 1996, cf. Sayer, 2000a). Interests are diverse and not as homogeneous as many management and economic models imply. Individuals, businesses and stakeholders hold a range of interests, some interests being morally based.

Figure 6.1 shows an idealised relationship between the market and the moral economy, two forces acting in opposite directions with perceived trade-offs between the two. In practice tensions do not mean pursuit of one is at the expense of the other as self-centred and trade-off

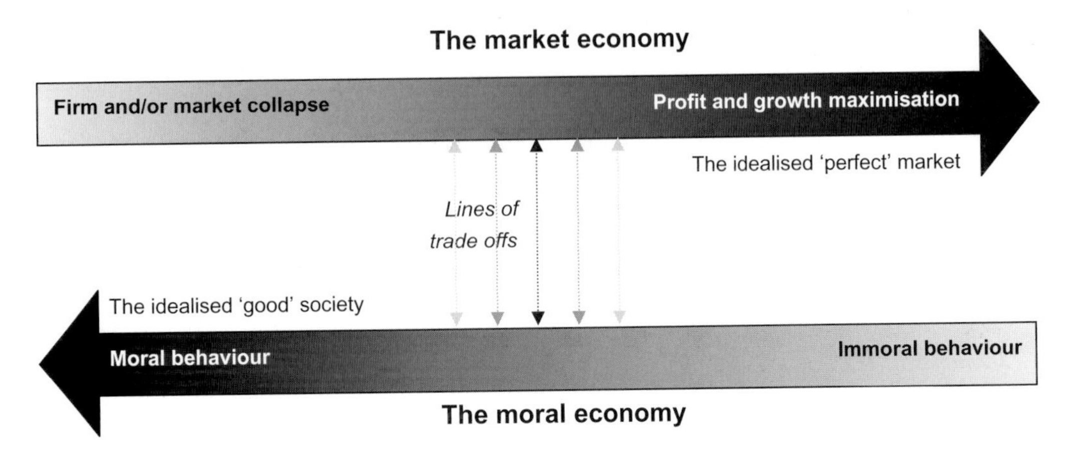

Figure 6.1 Idealised activity in the moral and market economies

assumptions of economics and game theory claim (Donaldson, 1989). A more realistic conception is presented in Figure 6.2. The 'continuums' are reformed as an arc, where they meet being zones of market and moral dysfunction. These extremes can invoke a broader crisis, sending a shock wave across the system, resulting in moral dysfunction that can affect the market, and ultimately inducing a market breakdown (see the narrow arc representing the affect shock wave in Figure 6.2).

It is the role of management to manage organisations and transactions in order that their respective organisations benefit, with the consequence that the extremes do not occur for the firms or for society. Yet, economic assumptions of selfishness and trade-off tend to deny these management roles and responsibilities. In practice, management is constantly making decisions informed by ethical issues (Hitt, 1990; Minkes *et al.*, 1999). Furthermore, management can proactively develop, nurture and maintain moral behaviour, whilst pursuing profit and growth – the zone of management action in Figure 6.3 – for example recognising the importance of how employees are treated (e.g. Pfeffer, 1994; Reichheld, 1996; Gummesson, 2001) for yielding competitive advantage.

The aggregation of behaviour, social exchanges in relationships and transactions builds the moral economy. The moral economy needs to be replenished from organisations in the market, building the moral stock. This is especially the case where transactions and exchanges are complex and long-term. However, there are benefits. 'Good' behaviour encourages further good behaviour (and *vice versa*), thus moral assets appreciate with use, in contrast with asset depreciation in the market economy. Developing trust is an investment in the business through the acquisition of social capital (Smyth and Pryke, 2006), and in transactions in the market economy.

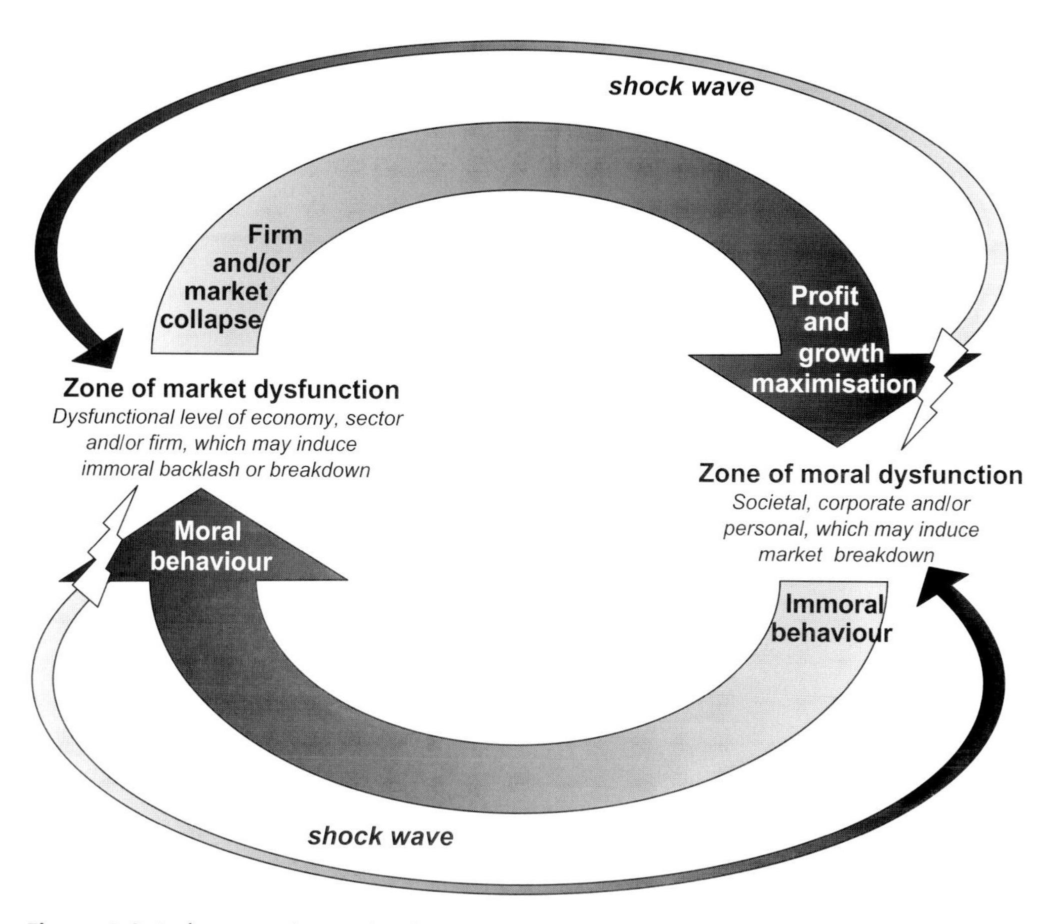

Figure 6.2 Dysfunction in the moral and market economies

Thus we can define the moral economy as the way in which morality in individual behaviour and the morality of social capital within a firm are brought to bear in market transactions to ensure that exchanges take place and that the market (supply) and use value (demand) of those exchanges are preserved at a minimum and potentially enhanced.

Methodological implications for trust

Trust is dependent upon specific conditions. The moral economy is contextual and helps frame market conditions too. Therefore, in researching trust it is important to select an epistemology (how we know what we know) and methodology that embrace context specificity, whilst

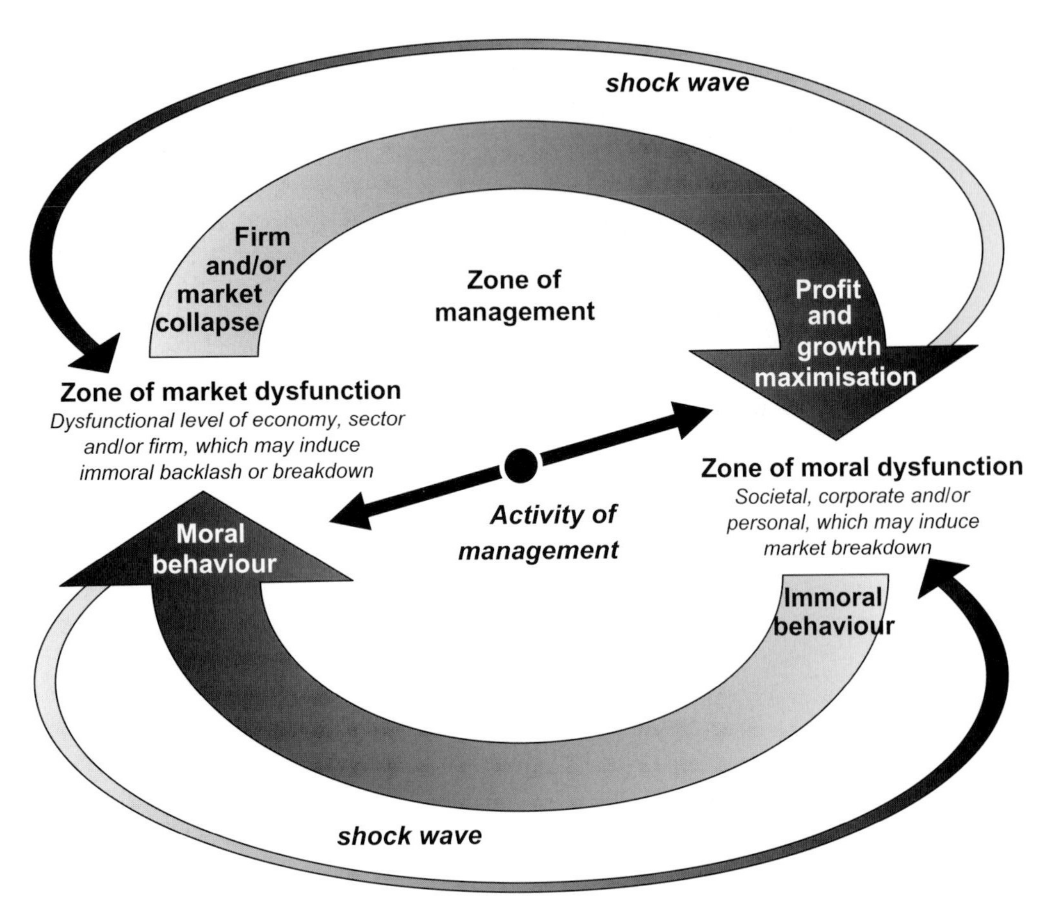

Figure 6.3 Management activity moving towards optimisation in the moral and market economies

preserving identification of general patterns (Smyth and Morris, 2007). Context is important to understand the dynamics and explore why particular outcomes occur, whilst the general patterns provide some guidance to management for future action, in this case developing trust. Figure 6.4 shows how philosophical issues inform epistemology and methodology for trust as the research object.

Pryke and Smyth (2006; see also Smyth *et al.*, 2006; Smyth and Morris, 2007) identified four paradigms for managing projects: *traditional* project management, *information processing*, *functional* management and the *relationship* approach. Trust fits closely within the *relationship* approach because it is through people and, hence, through relationships that value is added to products and services. The relationship paradigm accepts:

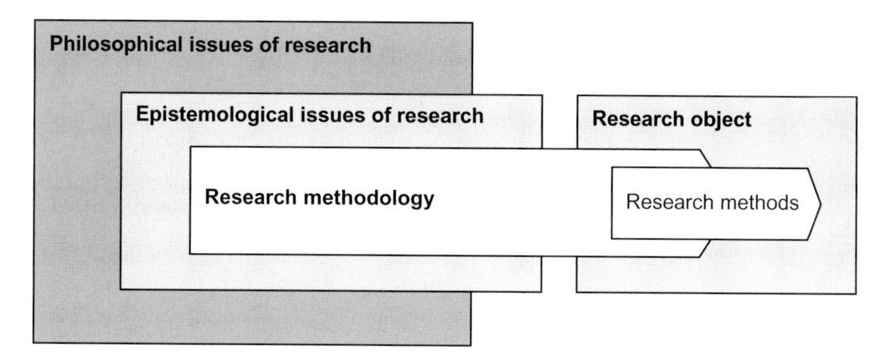

Figure 6.4 Applying epistemology, methodology and methods (adapted from Smyth and Morris, 2007)

- Management is contextual.
- Projects are contextual due to complexity and uncertainty.
- Relationships are always contextual:
  - Personal circumstances.
  - Team circumstances.
  - Organisational circumstances.
  - Market power and prevailing market conditions.

Critical realism provides one option (Bhaskar, 1975; Sayer, 1995, 2000b) for accounting how contextual conditions causally effect how trust will develop. It is compatible. Explanations arise from recognising the complexity of the real world, as opposed to an idealised one that simplifies causality to linear cause–effect explanations. Critical realism has two causal dimensions:

- The ontology or essence of the research object that embodies necessary causal powers and dispositions.
- The contextual conditions, in which the ontological powers may be released, each condition or set of circumstances carrying their own causal powers, such as management, project management, the market and moral economies.

These are shown schematically in Figure 6.5.

Ontology and dynamics of trust

Trust is held by an individual, but is not individualistic (Gilligan, 1997), for trust exists in relation to things and people. Even where a party is putting trust in a technology, social experience and the social effort involved in producing a technology are indirectly involved. Trust there-

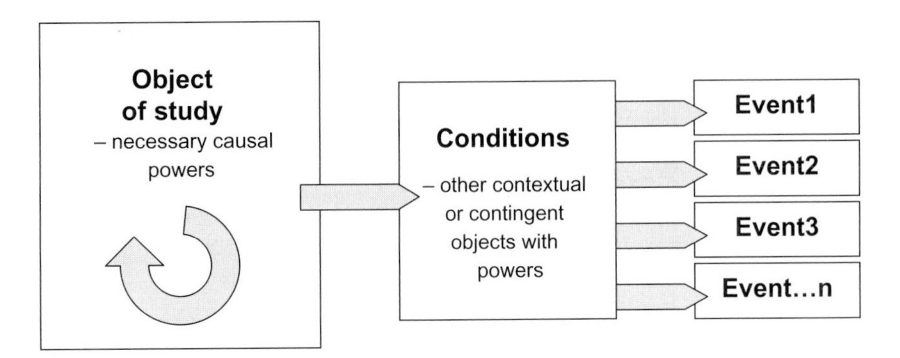

Figure 6.5 Realist causality (adapted from Smyth and Morris, 2007)

fore dynamically develops in relation to others, for evaluation of relationships is continual. As relationships change, trust is renegotiated; for example a sales-procurement relationship during pre-qualification, shortlisting and tendering changes on signing a formal contract as this imposes constraints on how relationships develop.

Trust is between people rather than organisations. Organisational norms, systems and procedures may affect trust between people, but 'trusting' an organisation is a surrogate, aggregate or mean (depending how parties evaluate organisations) of trust between the people with whom they have direct contact or a measure of brand strength and organisational reputation where there is no direct contact.

Trust is typically evaluated intuitively and frequently subconsciously, indeed it is the absence of trust that is more conscious (Dibben, 2004). Trust is intangible, therefore articulating the process of developing trust is important. Gilligan (1982) changed the way we understand psychological development. People are not islands of individualistic development, but develop in relationships with others. The inner being, the essence of the person, needs *voice* – to be listened to and heard by others – so that expression is given to who someone really is and the person grows. This important factor is necessary for personal growth; it is necessary for people to perform at their best and grow to their potential at work. When *voice* is extinguished by self or others, then relationships weaken and can break. A paradox exists where one party chooses to extinguish their voice for the 'sake' of preserving a relationship, yet the very act weakens the relationship, being seen where the other party tends to be selfish and dominant, with the less selfish party trying to preserve the relationship at their own expense. This is evident in blame cultures, such as many project environments, where even banter is used to 'score points' and establish hierarchies for blame (Smyth, 2000, 2004). Banter is paradoxically a healer of rifts, yet divides further. The real *voice* is silenced. Open communication is not the panacea. Open communication helps, perhaps reducing opportunism, yet it is usually expressed from a task-orientated perspective in projects (Handy, 1997). Absolute open communication means trust is not needed, and communication is highly aligned with extensive accountability, the dominant message being that others are not trusted (O'Neill, 2002; Smyth, 2003). Thus *voice* expresses the person, whereas most open communication in projects is at the surface level concerning tasks and accountability rather than people and their performance on projects through relationships.

People try hard, however, to maintain or invigorate relationships before breakdown. *Voice* provides opportunities to reconnect and (re)establish trust, yet a change in the nature of the relationship is necessary for voice to be heard (Gilligan, 1997). This is an instance of the importance of vulnerability to break into the situation and for managers to be facilitators. Managers need a map of trust development to help guide their actions. Despite scepticism of the possibility of *mapping the development of trust* (Luhmann, 1979), Dibben (2004) provides a starting point for mapping. In Table 6.2, typology 'A', one party (Y) is willing to rely upon the actions of or be vulnerable towards another party (Z), the minimal hope being that Z will not harm and preferably benefit Y in some way (Baier, 1994), which has parallels with self-interested trust. In practice, the emphasis between the trust framework elements may differ between relationships, involving the conditions of trust, social obligations of trust, characteristics of trust and components of trust, hence reality is complex. Indeed, the chronology may change too. Typologies 'B' and 'C' provide two further typologies based upon more positive behaviours in 'B' and negative behaviour in 'C'.

Context and circumstances for trust

The initial framework (Smyth, 2003) grouped elements as follows:

- *Concepts of trust*:
  - Characteristics of trust.
  - Components for trust.
- *Understanding the operation of trust:*
  - Conditions of trust.
  - Levels of trust.
  - Operational basis for trust.
- *Developing trust*:
  - Evidence of trust.
  - Trust in the marketplace.

The Egan Report (1998) repeatedly advocated, yet failed to explain, trust, or how it can be developed. The grouping above began to plug this gap, yet failed to fully account for the distinction between ontology and conditionality. The revised framework has two groups:

Table 6.2 Mapping the development of trust

A. Typology developed from Dibben (2004) (cf. self-interested trust)

- One party (Y) is willing to rely upon the actions of or be vulnerable towards another party (Z), the minimal hope being that Z will not harm and preferably benefit Y in some way (Baier, 1994)
- Y has a 'good' feeling or disposition towards Z on the first occasion of contact
- Y continues to have a 'good' feeling towards Z on several occasions, yielding an emotional investment in the relationship, and resulting in 'good' or positive behaviour towards Z
- Z responds with positive behaviour, yielding a subjective sense of identity with Z from Y
- Y continues to exhibit positive behaviour towards Z
- Z continues to exhibit positive behaviour towards Y, and the threshold of cooperation is reached
- Z and Y form expectations that the relationship will grow

B. Typology of exceeding expectations provided in 'A' (cf. towards socially orientated trust)

- Y is willing to rely upon the actions of or be vulnerable towards Z, the minimal hope being that Z will benefit Y in some way
- Y has a 'good' feeling towards Z on the first occasion of contact
- Y continues to have a 'good' feeling towards Z on several occasions, yielding an emotional investment in the relationship:
  - Specifically Y feels vindicated and affirmed in being willing to be vulnerable and trust Z
  - Y's trust turns to gratitude, e.g. increased client satisfaction if Y is the client, resulting in *very* 'good' or positive behaviour being exhibited towards Z
  - Y feeling able to trust Z to a greater extent, the initial evidence of trustworthiness being converted to confidence
  - Y may already feel an intuitive sense of identity with Z, yet looks for confirmation
- Z responds with positive behaviour, yielding confirmation of the subjective sense of identity with Z from Y: cooperative behaviour may be developing implicitly
- Y continues to exhibit positive behaviour towards Z:
  - Y's confidence in Z grows
  - Y's expectations for the relationship are raised to a higher level
- Z continues to exhibit positive behaviour towards Y, and explicit cooperation is achieved
- Z and Y form expectations that the relationship will grow

- Ontology of trust as the object of study:
  - Characteristics of trust.
  - Components for trust.
  - Social obligations of trust.
  - Conditions of trust.
- Conditions: context and circumstances for trust:
  - Levels of trust.
  - Operational basis for trust.
  - Trust in the marketplace.

Evidence of trust is methodologically located within events that feed back to affect context and therefore is no longer an element. Some further rationale for the framework categorisation is provided below:

Table 6.2 *Continued*

C. Typology of failing to meet expectations provided in 'A'

- Y is willing to rely upon the actions of or be vulnerable towards Z, the minimal hope being that Z will not harm Y in any way
- Y has a 'good' feeling towards Z on the first occasion of contact
- Z responds with behaviour that fails to confirm the initial 'good' feeling:
    - Y now feels unprotected in being vulnerable towards Z
    - Y has to decide whether to be vulnerable again, which may be linked to whether Y chooses to forgive themselves for setting their expectations too high (Baier, 1994), but if they fail to forgive themselves trust is eroded and the relationship is weakened
- Z responds with reasonable behaviour, yielding a subjective sense that there is scope for the relationship to develop
- Y continues to exhibit positive behaviour towards Z
- Z responds with negative behaviour towards Y:
    - Y chooses to forgive Z: *either* because Z took responsibility and acted reasonably, which in a market transaction meant Y paradoxically received less value than expected, yet gives an unconditional gift of forgiveness – less monetary value yet more moral investment; *or* because Z did not take responsibility and perhaps deliberately acted opportunistically too (Baier, 1994)
    - Alternatively Y fails to forgive themself and Z, yet maintains the relationship or retains the organisation to deliver another project, therefore trust is absent and Y will tend to blame Z for poor performance, even though responsibility may not entirely rest with them. The relationship changes and demands for greater accountability arise in a more adversarial way, *either* adversity may be motivated by defensiveness in order to be protected from blame from others within their respective employer organisations, *or* adversity may be motivated by attack in order to allocate blame to Z, if necessary 'ahead of time'
    - The consequence of an absence of trust and of failing to forgive is that Y is now harming Z (cf. Baier, 1994). The new expectation is that Z can no longer perform, even though they have been retained. Z is being 'set up' by Y perceptually to fail again. This new expectation ushers in a judgement, in fact a pre-judgement or prejudice towards Z
- Z and Y form expectations that the relationship will revert to a distant one or breakdown, depending upon whether there is an absence of trust or distrust

1. The importance of relationships:
    i. Trust is embedded and developed within relationships, so that while it is conceptually distinct as a conceptual category, hence relationships provide part of the context for development, trust cannot be divided from relationships in reality – a point made in psychology by Gilligan (1982) and philosophically by Gilligan (1997) and Baier (1994).
    ii. Long-term relationships are affected by the disposition of parties and the (un)willingness to develop trust; however, factors concerning personal circumstances and mood may affect short-term relationships.
    iii. Therefore, ontological factors of characteristics of trust and components of trust will be affected by contextual factors that cannot be easily distinguished through observation alone.

2. Ontology (see also Figure 6.6). Critical realism does not rely upon observation alone, recognising that all knowledge is conceptually and experientially informed. Social phenomena are generally complex, for example agency, thus relationships are particularly complex. These contextual dynamics act upon (the potential) for trust to induce particular events or outcomes, some of which can be grouped into a series of patterns whilst other events may be distinct. However, the conceptual elements help ontologically to show what trust is and what is possible, and therefore without these elements trust would become something less than it is or idealised into something it is not:
   i. The social obligations of trust are ontologically straightforward from a purely Kantian approach, yet the affect of the moral philosophy of care (Gilligan, 1982, 1997) and the more relativist society concerning tolerance and grace has meant that dignity and respect have become more contextual rather than purely a matter of the categorical imperative of duty. Dignity and respect are crucial to developing trust and inseparable in relationships, hence are important in forming the dynamic powers to develop trust. Yet the social obligations of trust are affected by the evidence parties rationally and intuitively pick up in context, and affected in cultural context and organisational norms by the extent of nurture and care.
   ii. Conditions of trust might be expected to be located in the grouping of *Conditions: context and circumstances for trust*; however, it is important to distinguish between the dynamics that contribute to the possibility of developing trust. In other words the conditions of trust help determine the possibility of trust in practice. Yet, contextual conditions will also act back to affect the conditions of trust, for example whether management choose to take a proactive role in encouraging the conditions of trust. A further conceptual reason for locating the conditions of trust within the ontological grouping is that there has been a series of studies trying to determine the 'antecedents of trust' (e.g. Wong *et al.*, 2000; Dirks and Ferrin, 2002; Bijlsma, 2003), typically in mechanistic cause–effect fashion, which have been contradictory and thus inconclusive (Smyth and Thompson, 2005). Locating the conditions of trust in juxtaposition to these studies, and indeed the social obligations of trust too, acts as a critique and challenge to these analyses of 'externalities'.
3. Context (see also Figure 6.6). Project management and the management of projects are two conceptual approaches that are both derived from context. The approach to managing projects adopted at any one time, arguably, has the potential to change the project context:
   i. Projects are managed individually or are managed through programme management (Pellegrinelli, 1997; Morris and

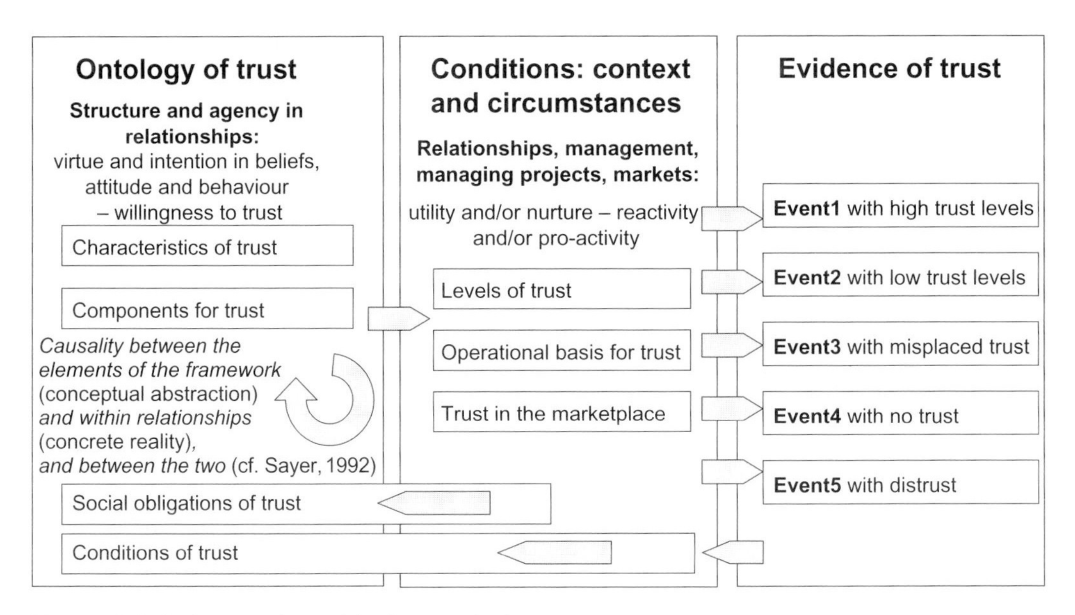

Figure 6.6 Realist causality and the framework of trust

Jamieson, 2004), or through a corporate relationship management strategy (Smyth, 2003; cf. Pryke and Smyth, 2006).

ii. Project teams pertain to organisations and are 'virtual' organisations that also span organisations (Cherns and Bryant, 1984) and thus contextually concern trust both within organisations and in the marketplace (see Figure 6.6).

Framework of trust

A revised framework of trust can now be proposed (see Table 6.1). The framework includes *social obligations of trust* as a new element. Each element is detailed below.

Characteristics of trust

The characteristics of trust are derived from the work of Lyons and Mehta (1997), which has bridged economic and social analyses of trust and integrated contractual aspects particularly from Sako (1992). This analysis also links the characteristics of trust to *relationship management* as a normative component for developing trust (cf. Storbacka *et al.*, 1994). Lyons and Mehta (1997) identified two characteristic types of trust:

1. *Self-interested trust* – a willingness to trust with minimal or no evidence for trust, from which it is estimated there is mutual short-term advantage to trust another party. Self-interested trust will continue providing there is no suspicion or mistrust. It can be summarised as being prepared to trust the other party until proved otherwise (Lyons and Mehta, 1997). The risk is small as is the initial reward. It is a classic 'win–win' situation, the motive being, 'What can the other party do for me?' (Smyth, 2003, 2005). Yet the possibility of building the relationship beyond the initial willingness to trust increases the reward to each party.
2. *Socially orientated trust* – generated through obligations in a social network of relationships, through reputation and advocacy and sustained through experience (Lyons and Mehta, 1997). Socially orientated trust features a preparedness to 'go the extra mile' for another trustworthy party, hence is sacrificial in terms of time and effort. Therefore, it not only develops from self-interested trust, but builds upon socially orientated trust both in one-to-one relations and in a broader network. Trust is developed with care through relationships, as each party must be aware of the willingness of the other party to be equally trustworthy. In business, investment is needed from both parties to transition to and maintain socially orientated trust. There must be a long-run return on the investment, yet each act of investment does not necessarily require short-term returns. In this sense trust is therefore unconditional, going beyond a simple short-term 'win–win'. It is fragile, partly because it can be lost quickly through opportunism (Lyons and Mehta, 1997), and partly because parties see the potential relationship value and investment as an asset subject to market risk. Long-term contracts and repeat business opportunities provide fertile conditions to develop and sustain socially orientated trust. It takes the definition of 'service' literally, the motive being: 'What can I do for the other party?'

Socially orientated trust requires some relationship history to develop (see the *map of the development of trust* above). However, transitioning from self-interested trust to socially orientated trust is contextually dependent; this is because:

- Unequal power relations between the parties at any time can affect socially orientated trust development, the more powerful party needing to avoid opportunism or deception, and preferably taking the initiative for transition to socially orientated trust.
- Different goals can inhibit development. While low level collaboration can aid the development of self-interested trust, identifying areas for cooperation facilitates socially orientated trust, hence avoiding opportunistic pursuit of goals at the expense of the other party. Shared goals favour development of socially orientated trust.

- Equal levels of investment by both parties in the relationship, management and technical requirements creates fertile ground for socially orientated trust to develop.
- Mutual respect and consistent responsibility taken for the relationship internally, as well as for the other party, help foster socially orientated trust.
- Mutual understanding of expectations, needs and likelihood to act in one way or another helps foster and maintain this trust.
- Expectations that both parties can secure a high and relatively equal level of benefits help foster and maintain socially orientated trust.

The development of socially orientated trust may require changes in organisational practice; for example 'accountancy driven' companies (where financial matters drive agendas rather than serving them) may require to shift towards 'core competency drivers' where collaborative capital gives rise to *relationship value* (the flow of income and profit from a relationship over time) as a key measure (cf. Storbacka *et al.*, 1994). For projects, the tactical and short-term task orientation will shift towards programme and portfolio management with a customer orientation.

Components for trust

Components for trust comprise a family of related (Edkins and Smyth, 2006a) and dynamic concepts (see Figure 6.7) associated with project team relationships:

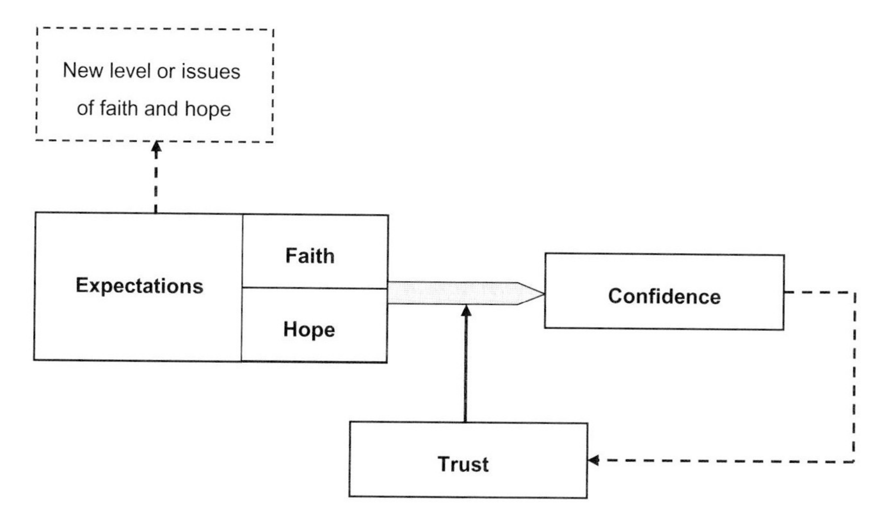

Figure 6.7 Components of trust dynamics (adapted from Edkins and Smyth, 2006a)

1. *Expectations* concerning the outcomes arising out of a working relationship with other parties:
   i. *Faith* in the unseen capabilities of other parties to perform – a belief, where there is scarce evidence.
   ii. *Hope* in the seen capabilities that other parties will perform – a belief, grounded in good reasoning to support the other party, derived from the achievement or expert opinions of others.
   iii. Trust, the willingness to be vulnerable, acts as a mediator, giving time and space in order to convert expectations into confidence, thus reducing perceived risk and uncertainty.
2. *Confidence* in others, a probability statement of successful outcomes derived from evidence of recent performance based upon direct and indirect experience.

Trust mediates between expectations and confidence, permitting confidence to grow and expectations to rise to a new level. This dynamic encourages parties to deepen their trust and hence move from self-interested trust to socially orientated trust. Continuing to meet expectations adds product and service value, expectations therefore rising beyond needs towards meeting desires. Supplier expectations are to increase market share, return on capital employed and profit margins. In this way, investments in trust as social capital can be expected to yield a return.

Social obligations of trust

The social obligations of trust comprise:

1. *Dignity* – to acknowledge the value of a party as a person that is based in *equality* as a 'human being' which cannot have a price attached to it.
2. *Respect* – to earn acknowledgement from another derived from behaviour and achievement as evidence of behaviour gleaned both from direct experience and through reputation, that is, in terms of *equity* as a 'human doing', which can have a price attached to it.

These aspects are linked through status and power in employment and the marketplace. Power confers dignity through status, and respect increases power and status, which affects market value, blurring distinctions through observation.

The definition of trust incorporated 'circumstances of contractual and social obligations'. Therefore the social obligations of trust provide an essential relationship factor, while contractual circumstances are context specific. Socially orientated trust is generated through obligations; therefore showing dignity and respect for others are important for developing socially orientated trust and providing initial evidence for trust to convert expectations to confidence (see the map of trust development above).

Conditions of trust

'Conditions of trust' considers the atmosphere and culture of trust derived from specific behaviours. Butler (1991) originally developed a 'conditions of trust inventory' of attributes and attitudes for consumer markets. Conditions of trust are particularly pertinent in business-to-business relationships. Hannah (1991), applying them to project teams, showed that performance derived from trust related to individuals rather than project factors. Thompson (2003) considered it in the client–contractor dyad, examining client perceptions. Thompson developed the conditions of trust inventory for construction, identifying behavioural intentions and abilities, presented in rank order of importance from the client perspective:

- *Behavioural intent* – inputs:
  1. Integrity (1)
  2. Receptivity (4)
  3. Loyalty (2)
  4. Discretion (5)
  5. Openness (3)
- *Behavioural ability* – outputs:
  1. Consistency (3)
  2. Promise fulfilment (2)
  3. Fairness (4)
  4. Competence (1)
  5. Availability (5)

The rank order differed in the consultant–client dyad (Smyth, 2005), the ranking shown in parentheses. These categories are not watertight conceptually for they are based upon perceptions, justifiable because conditional behaviour is informed by perception. The presence of conditions of trust reduces perceived risk. Although uncertainty and attendant risk are not objectively removed, the confidence in the other party increases the expectation that solutions will be forthcoming, even if they are unknown or unspecified at present (Smyth and Thompson, 2005). Therefore conditions of trust support the components of trust through confidence in relation to risk and provide evidence to encourage socially orientated trust.

Levels of trust

This is the first of the purely contextual elements and relates to organisational hierarchy. Lines of authority and the power inherently change relationships. Typically junior staff frequently share information and develop peer trust through openness, whereas senior staff frequently require more confidentiality in relationships with junior staff and therefore trust requirements are greater. Where the expectations of junior staff are aligned with the strategy of the organisation, then conformance of behaviour is more likely to develop and trust can too. Defining leadership is contentious and will affect the potential for trust to develop, especially socially orientated trust. Leadership influences organisational culture, taking two forms (Clark 1978):

1. *Communal norms* – revolve around love and protection, hence relate to the ethics of care (cf. Gilligan, 1982, 1997). Communal norms concern needs-based conduct of *equality*, which also relates to conditions of trust through fairness in particular, but also availability and receptivity. Positive communal norms will encourage socially orientated trust and management can take a proactive role through strategically developing a culture of nurture and care. The importance of internal relationships has been underscored by Reichheld (1996), demonstrating that dyadic relationships, the client–supplier interface, are no stronger than those within the organisation, for example loyalty. Communal norms display beliefs and values that go beyond the rules of governance within transaction cost analysis and game theory (Lyons and Mehta, 1997; Smyth, 2006).
2. *Exchange norms* – revolve around securing a good deal and therefore self-interest (Lyons and Mehta, 1997), are merit-based, and operate in the market according to the rules of *equity* (Kurtzberg and Medvec, 1999). Exchange norms apply where there is relational contracting and relationship management across the dyad, but only where an employment contract within an organisation in so far as the contract and working practices coalesce. On the other hand all relationships, even communal ones, have elements of negotiation and bargaining in them (Thompson, 2001), so boundaries are sometimes blurred in practice.

Operational basis for trust

The operational basis for trust is also contextual, depending upon an organisation's economic sector and strategies employed (Smyth, 2003). This is where trust is welded into or influences product and service value. There is a range of technical and managerial issues that are linked to the product or service that constitute operations. These are typically contextually specific; however, trust as a moral competency has broader application and relates to the discourse of this chapter. Trust as a moral competency (Smyth and Pryke, 2006) is potentially a core competency (cf. Hamel and Prahalad, 1996) and therefore a factor in the speed and depth of the formation of social capital (Smyth, 2003, 2006b; Pryke and

Smyth, 2006) – company assets derived from individuals and through relationships released to add product and services value. Morality relates to social capital in three ways:

- The moral economy is one origin of social capital.
- Development of social capital requires personal morality.
- Release of social capital into the firm and market depends upon moral behaviour amongst management and promoted by management.

Management may enhance the rate of social capital formation through the careful selection of the employees they hire. Effective management will facilitate the evolution or development of social capital through, for example:

- Management behaviours that encourage employees to take responsibility.
- Developing management systems and procedures that are conducive to the generation and growth of social capital.
- Employee rewards and incentive schemes.

Trust as social capital and as a core competency relies upon systems and procedures a firm puts in place in order to embed competencies into the firm, and also relies upon the proactive management of morality by employers to create and mobilise social capital.

Trust in the marketplace

The definition of trust incorporated 'circumstances of contractual and social obligations', hence contracts are contextually important. Contract markets are heterogeneous, trust operating in different ways, Fukuyama (1995) distinguishing between low- and high-trust societies. Communal norms and exchange norms are blurred by degrees in different cultures; for example exchange norms have greater communal content in many eastern nations. Within exchange norms there are also different approaches. Sako (1992) draws attention to:

- *Arm's-length contractual relationship* – characterised by low degree of interdependence, contract-orientated relationships and limited degree of trust, located in a range of behaviours from opportunism to self-interested trust, approximating to a pure conception of exchange norms.
- *Obligational contractual relationship* – characterised by heavy interdependence, long-term orientated relationships beyond contract duration and high degree of trust between the parties, located in a range of behaviours from self-interested trust to socially orientated trust,

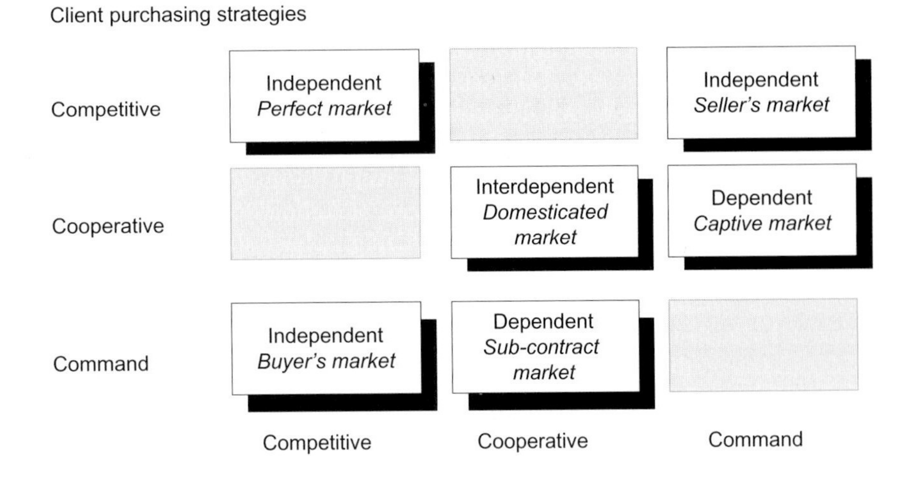

Figure 6.8 Classification of buyer–seller relationships (adapted from Campbell, 1995)

approximating to a conception of exchange norms with communal norms informing the relationships and moderating the exchange norms.

Trust ideally develops under mutuality with equality, whereas exchange is largely equity based in practice. Although project team members may share mutual aims such as the expectation of successful completion, totally aligned objectives are rare: clients require achievement of time–cost–quality/scope criteria with added service value; contractors desire profitable projects with repeat or referral business. Market power of the parties tends to further distort opportunities for mutuality. While power can shift during a contract (Gruneberg and Ive, 2000), the overall power relation is depicted in Figure 6.8. Trust is interpersonal, yet market relations and power frequently affect trust and behaviour.

However, where trust is proactively developed or is strong, then it acts to modify power and market relations. Figures 6.9 and 6.10 depict moves towards relational contracting in terms of top-down governance and the market on the one hand, and proactive, bottom-up management of behaviour on the other hand, for example through relationship management. As Baier (1994:105–106) points out:

Trust alters power positions, and both the position one is in without a given form of trust and the position one has within a relation of trust need to be considered before one can judge whether that form of trust is sensible and morally decent.

Client purchasing strategies

Competitive	Open tendering: traditional, D&B, turnkey	*Financial or supply chain leverage*	PFI & BOOT, target cost negotiated contracts
Cooperative	*Fragmented market*	Partnering contracts and framework agreements	Cost plus management fee Project management
Command	Selective tendering	Partnership	
	Competitive	Cooperative	Command

Contractor marketing strategies

Figure 6.9 Idealised project buyer–seller relationships

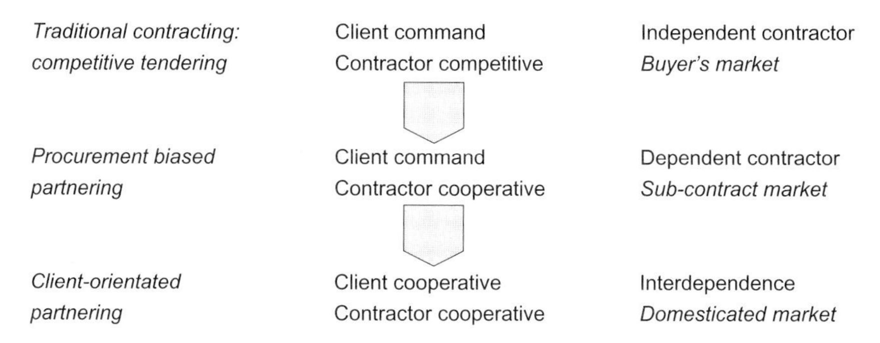

Figure 6.10 From a procurement bias to marketing–procurement balance

However, those who have more power both concerning the levels of trust and trust in the marketplace also carry more responsibility, and therefore, morally the onus is greater not to use opportunistic behaviour, a responsibility that much economic modelling, transaction cost analysis and game theory shirks.

Thus, power in the market affects trust but, conversely, trust across interpersonal relations affects the market. The balance of forces depends upon particular context, but also depends upon the moral economy present in the exchange and in the sectoral and broader economy at large. Organisations are therefore not passive recipients of market conditions but socially construct the market, trust being a very important part of this process.

Conclusion

This chapter has:

- Reviewed trust in relation to moral philosophy – virtue, intention, utility and nurture.
- Developed morality and trust in relation to the concept of the moral economy.
- Located trust within a methodology of critical realism.

The result has been a revision of a framework of trust:

- Regrouping elements into ontological and contextual categories.
- Inclusion of social obligations of trust.
- Exclusion of the evidence of trust, recognising that evidence arises through events.

This review and revision provides a more robust basis for developing trust conceptually and in practice as well as researching trust.

It has been argued that the role of trust is crucial in all exchanges and grows in importance for collaborative relationships. Trust grows through interpersonal behaviour. Trust is seen in the marketplace where it is proactively encouraged and managed to the extent that it is embedded in organisations and in evidence through aggregated behaviour of individuals in the organisation, and is in evidence indirectly through goodwill, branding and reputation. There are limits to the extent to which trust can grow through top-down changes to governance and market structures initiated by procurement drivers in construction. To this extent the analysis of collaborative relations generally, and trust in particular, through most economic models and game theory is correct, but this artificially limits the scope for collaboration. The argument has presented a proactive set of options for developing trust on the basis of a framework of trust, which unpacks the dynamics of trust, hence points to the areas in which management can act to develop trust and greater collaboration.

This argument can be perceived as prescriptive; however, such a position would miss the main point of the challenge, namely that the possibility is present, and it is to be seen whether clients, constructors and other supply chain members will take up the challenge or see the current practices of collaboration as the limit of action.

This analysis contributes to the relationship paradigm of managing projects, which has argued that the tools and techniques of traditional project management are only as useful as the hands they are in – through individuals and through relationships between individuals and teams in the project coalition.

References

Abolafia, M. (1996) *Making Markets: Opportunism and Restraint on Wall Street.* Harvard University Press, Cambridge.

Baier, A.C. (1994) *Moral Prejudices: Essays on Ethics.* Harvard Business Press, Cambridge.

Bhaskar, R. (1975) *A Realist Theory of Science.* Leeds Books, Leeds.

Bijlsma, K.M. (2003) Antecedents of trust in managers: a 'bottom up' approach. *Personnel Review*, 32, 638–664.

Butler, J.K. (1991) Toward understanding and measuring conditions of trust: evolution of conditions of trust inventory. *Journal of Management*, 17, 643–663.

Campbell, N. (1995) An interaction approach to organisational buying behaviour. In: Payne, A., Christopher, M., Clark, M. and Peck, H. (eds.) *Relationship Marketing for Competitive Advantage.* Butterworth Heinemann, Oxford.

Cherns, A.B. and Bryant, D.T. (1984) Studying the client's role in project management. *Construction Management and Economics*, 1, 177–184.

Clark, M.S. (1978) Reactions to a request for a benefit in communal and exchange relationships. *Dissertation Abstracts International*, 38(10-B), 5089–5090.

Cox, A. and Ireland, P. (2006) Relationship management theories and tools in project procurement. In: Pryke, S.D. and Smyth, H.J. (eds.) *The Management of Complex Projects: a Relationship Approach.* Blackwell, Oxford.

Currall, S.C. and Judge, T.A. (1995) Measuring trust between organisational boundary role persons. *Organisational Behaviour and Human Decision Processes*, 64(2), 151–170.

Dibben, M.R. (2004) Exploring the processual nature of trust and cooperation in organisations: a Whiteheadian analysis. *Philosophy of Management*, 4(1), 25–39.

Dirks, K.T. and Ferrin, D.L. (2002) Trust in leadership: meta-analytic findings and implications for research and practice. *Journal of Applied Psychology*, 87, 611–628.

Donaldson, J. (1989) *Key Issues in Business Ethics.* Academic Press, London.

Druskat, V.U. and Druskat, P. (2006) Applying emotional intelligence in project working. In: Pryke, S.D. and Smyth, H.J. (eds.) *The Management of Complex Projects: a Relationship Approach.* Blackwell, Oxford.

Edkins, A. and Smyth, H.J. (2006a) Contractual management in PPP projects: evaluation of legal versus relational contracting for service delivery. *ASCE Journal of Professional Issues in Engineering Education and Practice*, 132(1), 82–93.

Edkins, A. and Smyth, H.J. (2006b) The imperatives of trust: evaluations from the provision of 'full service' contracts. *Journal of Construction Procurement*, 12(2), 82–93.

Egan, Sir John (1998) *Rethinking Construction: The Report of the Construction Task Force.* Department of Trade and Industry, London.

Flores, F. and Solomon, R.C. (1998) Creating trust. *Business Ethics Quarterly*, 8(2), 205–232.

Ford, D., Gadde, L-E., Håkansson, H. and Snehota, I. (2003) *Managing Business Relationships.* Wiley, London.

Fukuyama, F. (1995) *Trust: The Social Virtues and the Creation of Prosperity.* Penguin Books, Harmondsworth.

Gambetta, D. (ed.) (1998) *Trust: Making and Breaking Cooperative Relations.* Basil Blackwell, Oxford.

Gilligan, C. (1982) *A Different Voice: Psychological Theory and Women's Development.* Harvard University Press, Boston.

Gilligan, C. (1997) *Voice and relationships: rethinking the foundations of ethics*, Presentation given 30th January, University of San Diego, http://ethics.acusd.edu/video/Gilligan/Lecture/Voice_and_Relationship.html.

Grönroos, C. (2000) *Service Management and Marketing.* John Wiley and Sons, London.

Gruneberg, S.L. and Ive, G.J. (2000) *The Economics of the Modern Construction Firm.* Macmillan, Basingstoke.

Gummesson, E. (2001) *Total Relationship Marketing.* Butterworth-Heinemann, Oxford.

Hamel, G. and Prahalad, C.K. (1996) *Competing for the Future.* Harvard Business School Press, Boston.

Handy, C.B. (1997) *Understanding Organizations.* Penguin, London.

Hannah, L. (1991) *Conditions of Trust in the Construction Industry and their Relevance to Project Success.* Research Implementation Report 91-01, Construction Industry Cooperative Alliance, Clemson University.

Hitt, W. (1990) *Ethics and Leadership: Putting Theory into Practice.* Battelle Press, Columbus.

Jones, C., Parker, M. and Bos, R.T. (2005) *For Business Ethics.* Routledge, Abingdon.

Kant, E. (1785) *Fundamental Principles of the Metaphysics of Morals.* Dover Publications, New York.

Kumaraswamy, M. and Rahman, M. (2006) Applying teamworking models to projects. In: Pryke, S.D. and Smyth, H.J. (eds.) *The Management of Complex Projects: a Relationship Approach.* Blackwell, Oxford.

Kurtzberg, T. and Medvec, V.H. (1999) Can we negotiate and still be friends? *Negotiation Journal*, 15(4), 355–362.

Luhmann, N. (1979) *Trust and Power.* John Wiley and Sons, London.

Luhmann, N. (1988) Familiarity, confidence, trust: problems and alternatives. In: Gambetta, D. (ed.) *Trust: Making and Breaking Cooperative Relations.* Basil Blackwell, Oxford.

Lyons, B. and Mehta, J. (1997) Contracts, opportunism and trust: self-interest and social orientation. *Cambridge Journal of Economics*, 21, 239–257.

Mayer, R.C., Davis, J.H. and Schoorman, F.D. (1995) An integrative model of organizational trust. *Academy of Management Review*, 20(3), 709–734.

Minkes, A., Small, M. and Chatterjee, S. (1999) Leadership and business ethics: does it matter? Implications for Management. *Journal of Business Ethics*, 20, 327–335.

Moorman, C., Deshpande, R. and Zaltman, G. (1993) Factors affecting trust in market research relationships. *Journal of Marketing*, 54, 81–101.

Morris, P.W.G. and Jamieson, A. (2004) *Translating Corporate Strategy into Project Strategy.* Project Management Institute, Newton Square, PA.
Norman, R. (1998) *The Moral Philosophers.* Oxford University Press, Oxford.
O'Neill, O. (2002) Lecture 1: Spreading Suspicion. *A Question of Trust.* Reith Lectures BBC 4, London.
Pellegrinelli, S. (1997) Programme management: organising project-based change. *International Journal of Project Management*, 15(3), 141–149.
Pfeffer, J. (1994) *Competitive Advantage through People.* Harvard School Business Press, Boston.
Polanyi, K. (1944) *The Great Transformation.* Beacon Press, New York.
Pryke, S.D. and Smyth, H.J. (2006) *The Management of Complex Projects: a Relationship Approach.* Blackwell, Oxford.
Rachels, J. (1985) *The Elements of Moral Philosophy.* McGraw-Hill, New York.
Reichheld, F. (1996) *The Loyalty Effect.* Harvard Business School Press, Boston.
Sako, M. (1992) *Prices, Quality and Trust: Inter-Firm Relations in Britain and Japan.* Cambridge University Press, New York.
Sayer, R.A. (1992) *Method in Social Science: a Realist Approach.* Routledge, London.
Sayer, R.A. (2000a) Markets, embeddedness and trust. *Research Symposium on Market Relations and Competition*, 4–5 May, Centre for Research Innovation and Competition, University of Manchester, Manchester.
Sayer, R.A. (2000b) *Realism and Social Science.* Sage, London.
Sayer, R.A. (2003) Restoring the moral dimension in social scientific accounts. *International Association for Critical Realism Annual Conference*, 15–17 August, Amsterdam.
Shaw, R.B. (1997) *Trust in the Balance: Building Successful Organizations on Results, Integrity, and Concern.* Josey-Bass Management Series, San Fransisco.
Smith, J.B. and Barclay, D.W. (1995) *Promoting Effective Selling Alliances: the Roles of Trust and Organizational Differences.* Technical Working Paper, Report No. 95-100, Marketing Science Institute, Cambridge, Massachusetts.
Smyth, H.J. (2003) *Developing Customer-Supplier Trust: a Conceptual Framework for Management in Project Working Environments.* CRMP Working Paper, http://www.crmp.net/papers/index.htm.
Smyth, H.J. (2004) Competencies for improving construction performance: theories and practice for developing capacity. *The International Journal of Construction Management*, 4(1), 41–56.
Smyth, H.J. (2005) Trust in the design team. *Architectural Engineering and Design Management*, 3(1), 193–205.
Smyth, H.J. (2006a) Measuring, developing and managing trust in relationships. In: Pryke, S.D. and Smyth, H.J. (eds.) *The Management of Complex Projects: a Relationship Approach.* Blackwell, Oxford.
Smyth, H.J. (2006b) The moral economy and operationalising trust. *Proceedings of RICS Cobra Conference*, 7–8 September, UCL, London.
Smyth, H.J., Morris, P.W.G. and Cooke-Davies, T. (2006) Understanding Project Management: philosophical and methodological issues. *Proceedings of Euram*, May 17–20, BI Management School, Oslo.
Smyth, H.J. and Edkins, A.J. (2007) Relationship management in the management of PFI/PPP Projects in the UK. *International Journal of Project Management*, 25(3), 232–240.

Smyth, H.J. and Morris, P.W.G. (2007) An epistemological evaluation of research into projects and their management: methodological issues. *International Journal of Project Management*, 25(4), 423–436.
Smyth, H.J. and Pryke, S.D. (2006) The moral economy and research on projects: neglect and relevance to social capital and competencies. *Proceedings of RICS Cobra Conference*, RICS Foundation, 7–8 September, UCL, London.
Smyth, H.J. and Thompson, N.J. (1999) Partnering and trust. *Proceedings of the CIB Symposium on Customer Satisfaction*, September, Cape Town.
Smyth, H.J. and Thompson, N.J. (2005) Managing conditions of trust within a framework of trust. Special Edition, *Journal of Construction Procurement*, 11(1), 4–18.
Storbacka, K., Strandvik, T. and Grönroos, C. (1994) Managing customer relationships for profit: the dynamics of relationship quality. *International Journal of Service Industry Management*, 5(5), 21–38.
Thompson, L. (2001) *The Mind and Heart of the Negotiator.* Prentice-Hall, New Jersey.
Thompson, N.J. (1996) Relationship marketing and advocacy.*Proceedings of the 1st National Construction Marketing Conference*, July 4, Centre for Construction Marketing in association with the Chartered Institute of Marketing Construction Industry Group, Oxford, Oxford Brookes University.
Thompson, N.J. (1998) Can clients trust contractors? Conditional, attitudinal and normative influences on clients' behaviour. *Proceedings of the 3rd National Construction Marketing Conference*, July 9, Centre for Construction Marketing in association with the Chartered Institute of Marketing Construction Industry Group, Oxford, Oxford Brookes University.
Thompson, N.J. (2003) *Relationship Marketing and Client Trust Toward Contractors within the Large Private Building Sector of the UK Construction Industry.* PhD, Oxford Brookes University.
Wong, E.S., Then, D. and Skitmore, M. (2000) Antecedents of trust in intra-organisational relationships within three Singapore public sector construction project management agencies. *Construction Management and Economics*, 18, 797–806.

Appendix 1

A sample of 18 articles on trust with over 300 citations on Google Scholar (15 January 2007) in the category of 'Business, Administration, Finance and Economics' have been drawn upon, of which 17 articles were reviewed, positivist methodologies being used in 14 (82%) articles, 15 if the empiricist tradition is included; 12 (80%) articles used quantitative tools and techniques of the 15 articles that undertook fieldwork. Yet none of the authors specified the methodology, and reasons for selection, nor reviewed the methodological approach in the light of the evidence, which acted as the template of influence for studies in construction collaboration (Table 6.3).

Table 6.3 Review of most cited management articles on trust

Authors	Article details	Google scholar citations	Methodology	Method	Post-methodology review
Morgan, R.M. and Hunt, S.D. (1994)	The commitment-trust theory of relationship marketing. *Journal of Marketing*, 58, 20–38.	1817	Positivist	Quantitative	No
Mayer, R.C., Davies, J.H. and Schoorman, F.D. (1995)	An integrative model of organizational trust. *The Academy of Management Review*, 20, 709–734.	1131	Positivist	Qualitative	No
Gulati, R. (1995)	Does familiarity breed trust? The implications of repeated ties for contractual choice in alliances. *The Academy of Management Journal*, 38, 85–112.	732	Positivist	Quantitative	No
Doney, P.M. and Cannon, J.P. (1997)	An examination of the nature of trust in buyer-seller relationships. *Journal of Marketing*, 61: 35–51.	675	Positivist	Quantitative	No
Rousseau, D.M., Sitkin, S.B., Burt, R.S. and Camerer, C. (1998)	Not so different after all: a cross-discipline view of trust. *The Academy of Management Journal*, 23, 393–404.	568	Interpretative	Qualitative	Yes*
McAllister, D.J. (1995)	Affect- and cognition-based trust as foundations for interpersonal cooperation in organizations. *The Academy of Management Journal*, 38, 24–59.	551	Positivist	Quantitative	No
Jarvenpaa, S.L. and Leidner, D.E. (1999)	Communication and trust in global virtual teams. *Organization Science*, 10, 791–815.	485	Empiricist	Quantitative	No
Moorman, C., Zaltman, G, and Deshpande, R. (1992)	Relationships between providers and users of market research: the dynamics of trust within and between organizations. *Journal of Marketing Research*, 29, 314–328.	427	Positivist	Quantitative	No
Garbarino, E. and Johnson, M.S. (1999)	The different roles of satisfaction, trust, and commitment in customer relationships. *Journal of Marketing*, 63, 70–87.	382	Positivist	Quantitative	No

(Continued)

Table 6.3 *Continued*

Authors	Article details	Google scholar citations	Methodology	Method	Post-methodology review
Bertrand, M., Duflo, E. and Mullainathan S. (2004)	How much should we trust differences-in-differences estimates? *The Quarterly Journal of Economics*, February, 248–275.	380	Positivist	Quantitative	No
Glaeser, E.L., Laibson, D.I., Scheinkman, J.A. and Soutter, C.L. (2000)	Measuring trust. *The Quarterly Journal of Economics*, 115, 811–846.	370	Positivist*	Quantitative	No
Das, T.K and Teng, B-S. (1998)	Between trust and control: developing confidence in partner cooperation in alliances. *Academy of Management Review*, 23, 491–512.	366	Positivist**	n/a	No
Moorman, C., Deshpande, R. and Zaltman, G.I. (1993)	Factors affecting trust in market research relationships. *Journal of Marketing*, 57, 81–101.	366	Positivist	Quantitative	No
Zaheer, A., McEvily, W. and Perrone, V. (1998)	Does Trust Matter? Exploring the effects of interorganizational and interpersonal trust on performance. *Organizational Science*, 9, 141–159.	336	Positivist	Quantitative	No
McKnight, D.H., Cummings, L.L. and Chervany, N.L. (1998)	Initial trust formation in new organizational relationships. *Academy of Management Review*, 23, 473–490.	326	Positivist**	n/a	No
Hosmer, L.T. (1995)	Trust: the connecting link between organizational theory and philosophical ethics. *Academy of Management Review*, 20, 379–403.	325	u/k**	Qualitative	No
Zak, P.J. and Knack, S. (2001)	Trust and growth. *The Economic Journal*, 111, 295–321.	323	Positivist	Quantitative	No

* Methodological aspects are touched upon regarding the work of others rather than a review of their own epistemology and related methodologies.
** Articles that are essentially analysis or synthesis derived from Literature Reviews.

Section III

Collaborative Relationships and Networks

In this section tendencies and conceptual potential in moving further towards collaborative relationships embedded in networks are considered. The section describes how frameworks and programmes dovetail and develop into networks. Networks as social relationships go beyond the boundary of firms and organisations.

In **Chapter 7** infrastructure lifecycles and disaster mitigation are considered by Haigh, Amaratunga, Keraminiyage and Pathirage, particularly the role of government, NGOs and the construction sector in contingency planning for disasters. This addresses collaboration at the conceptual level of policy and administrative networks. Volker addresses early design management in architecture in **Chapter 8**. Selecting partners for and value judgment in design decision making and development provide the empirical focus, and the chapter considers how the project manager is appointed in the design phase with strategic objectives in mind. In **Chapter 9** London and Chen explore the role of government. Policy and policy implementation gaps are considered concerning interdepartmental and agency decision responsibilities that contribute to project outcomes. **Chapter 10**, by Amaratunga, Shanmugam, Haigh and Baldry, focuses upon the construction workforce, particularly the representation of women in construction and especially amongst the professions, drawing upon the experience of women in the medical profession as a comparison.

7 Disaster mitigation through collaborative knowledge-sharing practices

Richard Haigh, Dilanthi Amaratunga, Kaushal Keraminiyage and Chaminda Pathirage

Background

The number of reported disasters has increased steadily over the past century and risen very sharply during the past decade according to the World Disasters Report 2005. An average of 354 disasters of natural origin occurred per year in the period 1991–1999. From 2000–2004, this rose to an average of 728 per year. The Asia–Pacific region has experienced the greatest loss of life in absolute terms and in proportion to the population, due to earthquakes, floods and tropical cyclones. In economic terms, the World Disasters Report 2002 assesses the average estimated damage due to natural disasters at US$69 billion. Asia experiences the highest reported losses but those in Europe are considerably greater than those in Africa. This reflects the high value of the infrastructure and assets at risk.

The Asian region is highly prone to natural hazards, but a hazard in itself does not necessarily lead to a disaster that brings about loss of lives, property and employment, and damage to the physical infrastructure and the environment (de Guzman, 2002). Natural hazards like earthquakes, however intense, inevitable or unpredictable, translate into disasters only to the extent that the population is unprepared to respond, unable to cope, and, consequently, severely affected. An earthquake will cause little damage if it takes place in an empty desert. It may also cause little damage if it takes place where people can afford to be well protected. Hence, a natural event only causes serious damage when it affects an area where the people are at risk and poorly protected. Disasters occur when these two factors are brought together.

The new millennium has seen a series of events – including the Bam earthquake in Iran (2003), the Indian Ocean tsunami (2004) and the terrorist events in America (2001), Madrid (2004) and London (2005) – that have increased the degree of uncertainty faced by policy makers, challenged emergency arrangements and raised issues regarding their appropriateness. In light of these events, the terms *resilience* and *disruptive challenges* have been adopted by many policy makers in an attempt to describe the way in which they would like to reduce a nation's susceptibility to major incidents of all kinds by reducing their probability of occurring and their likely effects, and by building institutions and structures in such a way as to minimise any possible effects of disruption upon them (Coles, 2004; Civil Contingencies Secretariat, 2004).

The concept of resilience has arisen from an amalgamation of historic developments in the disaster planning process. It has a focus on disaster and addresses the ability of the community to recover following the impact of a disastrous event (Fox, 2002). Douglas and Wildavsky (1982) define resilience from the perspective of risk as 'the capacity to use change to better cope with the unknown: it is learning to bounce back' and emphasise that 'resilience stresses variability'. More recently but in a similar vein, Dynes (2003) associates resilience with a sense of emergent behaviour which is improvised and adaptive, while Kendra and Wachtendorf (2003) draw parallels with the creative actions of organisations in the aftermath of disasters. Creativity, they argue, is a vital element in emergency response and emphasis should be placed on better preparation and training employees, to enhance creativity at all levels of responding organisations.

It is evident that the development of resilience to disasters is a complex problem and multi-disciplinary in nature. This chapter explores the role of construction professionals in the resilience agenda. It considers the challenges associated with greater collaboration and the need for an improved knowledge base in the hope of a more well thought out and successful approach to the problem. The chapter begins by considering existing national policies in resilience and recovery, and considers the construction professional's role within them. A more expansive view of the lifecycle of infrastructure projects is offered, one that extends beyond the traditional cycle of feasibility analysis, planning, design, construction, operation, maintenance and divestiture. This revised lifecycle should acknowledge the multi-disciplinary and collaborative nature of the resilience agenda, and encompass the construction professional's ability to anticipate and respond to unexpected events that damage or destroy an infrastructure project. The chapter concludes by examining the need to raise knowledge and awareness among construction researchers and practitioners to identify those aspects of resilience that will benefit from the involvement of construction professionals. An enhanced knowledge base is proposed, that enables the sharing of good practices, and improved collaboration through the development of a worldwide network of trained professionals ready to join recovery and reconstruction teams working with local people.

National policy in resilience and recovery

Resilience and recovery require a concerted approach that will support the foundations of community sustainability and capacity building and which will eventually reduce risks and vulnerabilities to future disasters (Rotimi *et al.*, 2006). The rational starting point is the setting up of an institutional infrastructure for emergency management, which will formulate public policies for preparedness, response and recovery. These recovery policies should then be integrated into other emergency management areas as well as policies of sustainability and community capacity building (Coghlan, 2004).

In the UK, the resilience policy agenda is being driven by the Civil Contingencies Secretariat (CCS), which was set up to improve the UK's resilience against disruptive challenges through working with others to anticipate, assess, prevent, prepare, respond and recover. The CCS has a two-stranded approach focusing on a Capabilities Programme (Cabinet Office, 2003) and the Civil Contingencies Bill 2004. *Capability* is a military term which is intended to be inclusive of personnel, equipment and training and such matters as plans, doctrine and the concept of operations (Coles, 2004). Capabilities are embedded in systems, routines, mechanisms and practices, and can only be improved by making such features more responsive to new needs. Capability is a forward-looking view: it predicts the outcome before a process has taken place. In contrast, performance is backward looking and consists of the results achieved after completion, thus providing historic data (Haigh, 2003). The Capabilities Programme is an audit of current infrastructure and resources extending to the full range of contingencies likely to face the UK in the first decade of the twenty-first century. The aim is to ensure that a robust infrastructure of response is in place to deal rapidly, effectively and flexibly with the consequences of civil devastation and widespread disaster inflicted as a result of conventional or non-conventional disruptive activity (Civil Contingencies Secretariat, 2004).

The CCS stresses the need for an integrated planning perspective in order to achieve resilience. Specifically, any planning system should include all levels of Government as well as the public, private and voluntary sectors. It defines resilience as including a planning process based on partnerships, the sharing of best practice and systems that are developed and tested to cover the full range of potential, disruptive hazards. This type of multi-disciplinary approach is widely supported (Fox, 2002; Tobin and Whiteford, 2002). However, Coles (2004) notes the complex and confusing picture of departmental domains and responsibilities of the Capabilities Programme, and argues that it is 'a clear example of how to further extend complexity in what is already a complex, tightly coupled system of interactions and interdependencies'. This complexity also appears to be at odds with improvisation and creativity of resilience, which Dynes (2003) contrasts with the emergency response organisations' command and control structure that destroys flexibility and innovation.

In a similar vein, the Ministry of Civil Defence and Emergency Management (MCDEM) in New Zealand encourages a holistic approach to the issue of recovery planning and believes this will be most effective if it is integrated with reduction, readiness and response. The definition of recovery encapsulates the expectations of recovery as 'the coordinated efforts and processes to effect the immediate, medium and long-term holistic regeneration of a community following a disaster' (MCDEM, 2004). Recovery is delivered through a continuum of central, regional, community and personal structures. The MCDEM, together with cluster groups of agencies, coordinates planning at the central level. Regional and territorial authorities are encouraged to produce group plans that will suit peculiar conditions of their local areas. Other discussion documents produced at the national level like *Focus on Recovery: A Holistic Framework for Recovery* and *Recovery Planning*, both released in 2004, give context to recovery planning while the Civil Defence and Emergency Management Act (CDEMA) 2002 provides the legislation and the foundations for the New Zealand Civil Defence and Emergency Management (CDEM) environment (Rotimi *et al.*, 2006).

Elsewhere, and in developing countries in particular, recovery planning appears less defined and, when tested, is often found to be insufficient or ineffective. Jigyasu (2004) describes an increase in vulnerability of local communities after the Latur 1993 earthquake in India, where sustainable recovery interventions were poorly planned and implemented. The national and international humanitarian response to the 2004 tsunami varied in each of the countries given differences in the area it covered, political context, impact as percentage of economy, culture, existing relationships to the donor nations and access to resources. In India and Sri Lanka, Rex (2006) found they naturally differed in all of the above, although recognised that within both countries, there was not one homogeneous culture but numerous, with often confusing and contentious, sub-cultures to be dealt with. In Sri Lanka, Rex found this to be more obvious due to awareness brought by the civil war, but in India it came as a surprise that there was so much variation due to combinations of geography, religion, caste, livelihood and language. In general however, Rex found that relief and reconstruction efforts in all locations were affected by:

> *Sudden availability of great quantities of money with minimal systems for spending; increased visibility and focus on corruption; wide geographical spread of relief operations making logistics and communications difficult; and, competition for resources.* (Rex, 2006:56)

Construction's role in resilience and recovery

The recovery role of construction from both natural and human disasters is well documented. In particular, post-disaster reconstruction has been the subject of a significant body of research (for example Jigyasu, 2002; Karim, 2004; Lizarralde and Boucher, 2004; Nikhileswarananda, 2004; Young, 2004) with particular emphasis on developing countries that are less able to deal with the causes and impacts of disasters. The importance of improving the construction industries of developing nations is widely recognised, highlighting a need to equip them to manage recovery (Ofori, 2002). Construction is typically engaged in a range of critical activities: temporary shelter before and after the disaster; restoration of public services such as hospitals, schools, water supply, power, communications, and environmental infrastructure, and state administration; securing income earning opportunities for vulnerable people in the affected areas (World Bank, 2001). Similarly, disaster planners have begun to realise the link between disaster and development (Fox, 2002) – a large and well established field relating to social, economic and, significantly, from a construction perspective, physical aspects of society.

Although more robust construction in and of itself will not eliminate the consequences of disruptive events, there is widespread recognition that the engineering community has a valuable role to play in finding and promoting rational, balanced solutions to what remains an unbounded threat (Sevin and Little, 1998). There has been considerable research aimed at developing knowledge that will enable the construction of a generation of buildings that are more resilient and safer, for example, through reduction of injury-inducing blast debris, the development of glazing materials that do not contribute to the explosion-induced projectiles and have enhanced security application, as well as the integration of site and structure in a manner that minimises the opportunity for attackers to approach or enter a building (Levy and Salvadori, 1992; National Research Council, 1995; Mallonee *et al.*, 1996; The President's Commission on Critical Infrastructure Protection, 1997).

A more expansive construction lifecycle

There is growing recognition that the construction industry has a much broader role to anticipate, assess, prevent, prepare for, respond to and recover from disruptive challenges. The process of disaster management is commonly visualised as a two-phase cycle, with post-disaster recovery informing pre-disaster risk reduction, and vice versa. The disaster management cycle illustrates the ongoing process by which governments, businesses and civil society plan for and reduce the impact of disasters, react during and immediately following a disaster and take steps to recover after a disaster has occurred (Warfield, 2004). The significance of this concept is its ability to promote the holistic approach to disaster management as well as to demonstrate the relationship between disasters and development (de Guzman, 2002). Recovery and reconstruction are commonly identified within the post-disaster phase, the period that immediately follows after the occurrence of the disaster. Once a disaster has taken place, the first concern is effective *recovery*; helping all those affected to recover from the immediate effects of the disaster. *Reconstruction* involves helping to restore the basic infrastructure and services which the people need so that they can return to the pattern of life which they had before the disaster (Davis, 2005). The importance of the 'transitional phase', linking immediate recovery and long-term reconstruction, is also stressed by a number of publications (de Guzman, 2002; Lloyd-Jones, 2006). With the recovery of social institutions, the economy and major infrastructure, efforts may shift to longer-term recovery and reconstruction.

The pre-disaster phase of the disaster management cycle includes both *mitigation* and *preparedness*. The RICS (Lloyd-Jones, 2006) propose that disaster mitigation refers to any structural and non-structural measure undertaken to limit the adverse impacts of natural hazards, environmental degradation and technological hazards. Mitigation measures may eliminate or reduce the probability of disaster occurrence, or reduce the effects of unavoidable disasters. These measures may include building codes; vulnerability analyses updates; zoning and land use management; building use regulations and safety codes; preventive health care; and public education (Warfield, 2004). Mitigation seeks to eliminate the risk of future disasters by effective sharing of lessons learned through preparedness planning.

Peña-Mora (2005) suggests construction managers have a key role to play because they are involved in the construction of the infrastructure, and therefore should also be involved when an event destroys that infrastructure. He highlights construction management skill in getting equipment, scheduling a set of activities to accomplish a task and knowing how to manage those activities, which can be very valuable when an extreme event occurs. Moreover, he stresses that construction engineers possess valuable information about their projects, and that information can be critical in disaster preparedness, as well as response and recovery. The information they posses, he argues, may be the difference between life and death. Similarly, Lloyd-Jones (2006) concluded that chartered surveyors, with appropriate training, have key roles to play during all disaster phases, from preparedness to immediate relief, traditional recovery and long-term reconstruction.

Sevin and Little (1998) suggest that computerised building plans, structural analysis programmes and damage assessment models may all facilitate rapid rescue and recovery of victims in the aftermath of an event, and that these all require the active involvement of the construction professions. They also suggest that the construction professions are in the best position to frame the discussion of the cost–benefit trade-offs that occur in the risk management process, for example the need for risk avoidance against the cost of implementing safety strategies.

The multi-disciplinary and integrated planning approach that is being adopted by the CCS and MCDEM would suggest a clear role for the construction industry in contributing to national resilience agendas. At present, the full potential for construction to contribute to such initiatives requires further clarification beyond traditional design and reconstruction activities.

In developing countries, construction's role is equally important but also different. Following a major disaster, such as the 2004 tsunami, both human and material resources were over-extended which slowed progress and drove up prices. The internal difference in resources derives from the situation that in a typical development programme, construction is very basic and in developing countries very little expertise is required at the scale a programme operates. However, when they must increase the capacity in quantity and speed, more experienced and skilled resources are required to actually manage projects. In addition to construction expertise, basic and complex project management experience is also essential for projects of varying scale. For both of these skill-sets, smaller developing countries may not have people with this large-scale time-critical management experience and expatriates with these skills frequently are inexperienced in understanding the local environment adequately (Rex, 2006).

Enhancing the construction knowledge base in resilience and recovery

Peña-Mora (2005) suggests the need for a more expansive view of the lifecycle of infrastructure projects, one that extends beyond the traditional cycle of feasibility analysis, planning, design, construction, operation, maintenance and divestiture. This revised lifecycle should encompass the construction professional's ability to anticipate and respond to unexpected events that damage or destroy an infrastructure project – from earthquakes to terrorist attacks – and reflect construction's ongoing responsibility toward an infrastructure's users. He refers to this concept as the *enhanced management and sustainability cycle for infrastructure with uncertain life span* or *enhanced lifecycle*.

Designing policy for future events crucially depends upon the degree of certainty or uncertainty faced by the architects of such policy. It is evident that the disruptive challenges of recent years have served as a 'wake-up call' for governments and have increased the degree of uncertainty faced by policy makers. A resilience agenda suggests a need for creativity, improvisation and adaptation, but it is unclear what role construction professionals will have in the 'complex and confusing picture of departmental domains and responsibilities', that national recovery programmes frequently offer (Coles, 2004:156). Creativity has been identified as a key component of resilience, and therefore the full extent of construction's potential contribution is likely to be best identified by construction professionals themselves, who can identify a more expansive view of the lifecycle of infrastructure projects that encompass the construction professional's ability to anticipate and respond to unexpected events that damage or destroy an infrastructure project. Knowledge and awareness among construction researchers and practitioners will identify those aspects of resilience that will benefit from the involvement of construction professionals and, where appropriate, stimulate ideas for further research. Lloyd-Jones (2006) concludes that the built environment's professional bodies, such as the RICS and CIOB, should work together to promote the development of a worldwide network of trained professionals ready to join recovery and reconstruction teams working with local people. This will be most effective if the various construction-related disciplines – conceivably through their respective professional bodies – collaborate efficiently towards the common goal of increased resilience, which will involve mutual learning to understand the considerable challenges to be faced and identifying how best to coordinate their response.

At present however, this knowledge base appears fragmented and poorly integrated with resilience and recovery policies. Eighteen months on, most of the affected counties had yet to recover from the devastation caused by the Indian Ocean tsunami of December 2004. Sri Lanka is no exception; overall the tsunami affected two-thirds of the coastline of Sri Lanka. It also resulted in the destruction of more than 100 000 houses (UNEP, 2005). The destruction of houses also resulted in discontinuance of several livelihoods such as fishing, farming, tourism and handicrafts-related activities. The UNEP report highlights the context in which the current post-tsunami rehabilitation is operating. Among the most important factors is the pre-existence of very high densities of unplanned settlements in the southern part of Sri Lanka with the majority of construction not observing some of the critical building standards. To add to this, the post-tsunami rehabilitation operation has been affected due to weak local government institutions with poor response capacities to address the needs of such a magnitude. This is mainly because, before the tsunami, Sri Lanka was known to be a safe haven where natural disasters scarcely occurred except for occasional floods and landslides during the rainy seasons.

In addition to commercial and non-commercial property damage, the number of deaths apportioned to the Indian Ocean tsunami is estimated to be in excess of 250 000, with at least 40 000 of those in Sri Lanka (BBC, 2005). A lack of awareness has been identified as a major reason behind the huge loss of life (Banerjee, 2005). Indeed, the term *tsunami* was heard by most of the ordinary Sri Lankans only after this devastation. Many of the direct victims were affected at the scene as they were curiously observing the pre-warnings of tsunami, without knowing the nature or the scale of the disaster to come. Both awareness and preventive steps are needed to prevent such huge loss of human life in future. For prevention, in the future, the first step is to sensitise people at large and create awareness through different media and text on various natural hazards, including the tsunami, and the preventive measures to be adopted (Banerjee, 2005). The problem continues beyond the pre-disaster stage into recovery, where Sri Lanka has again demonstrated the need for proper information and knowledge dissemination, as this has often been highlighted as the reason behind unsuccessful post-tsunami recovery activities. This can be achieved in part through construction professionals collaborating with community-based organisations. For example, following the tsunami, the Sri Lankan Institute of Civil Engineers worked with some communities to develop simple construction guidelines and standards to aid community-based reconstruction efforts. Such collaborative practices need to be far more widespread, and then disseminated both nationally and internationally. A lack of prior knowledge and proper points of reference have made most of the recovery plans guessing games, eventually failing without adding appropriate values to the recovery attempts (Banerjee, 2005).

Evidence from tsunami-affected regions suggests that there needs to be a conscious effort made towards managing knowledge at a national, provincial and sub-provincial level (Mohanty *et al.*, 2006). Knowledge can be differentiated between explicit, tacit and implicit forms of knowledge. Explicit knowledge is that which is stated in detail and leaves nothing merely implied. It is termed *codified* or *formal* knowledge because it can be recorded. Tacit knowledge is that which is understood, implied and exists without being stated, mainly grounded in individuals. In an organisational context, knowledge management is about applying the collective knowledge of the entire workforce to achieve specific organisational goals and facilitating the process by which knowledge is created, shared and utilised (Nonaka and Takeuchi, 1995). However, within a disaster management context, knowledge management is all about getting the right knowledge, in the right place, at the right time (Mohanty *et al.*, 2006). If a strategic approach is to be adopted to achieve disaster management objectives, knowledge management will play a valuable role in leveraging existing knowledge and converting new knowledge into action.

Disaster management knowledge can be identified at three different levels: institutional, group and individual, in the forms of both tacit (primarily) and explicit knowledge. The linkages among all agencies working on disaster management need to be strengthened in order to derive the regional best practices and coping mechanisms (Lloyd-Jones, 2006). In order to enhance the information sharing and management of the knowledge generated in institutions, it is important to closely knit these organisations and institutions and, moreover, groups and people working within them (UNDP, 2005). As alluded to previously, there are many gaps that may be bridged by appropriate use of construction professionals' skills, but access to these by local organisations that are on the front line of the mitigation, recovery and reconstruction efforts are highly constrained by a lack of recognition and understanding of their existence. Local organisations must be able to collaborate effectively with construction professionals to address the significant infrastructure and built environment-related challenges when developing resilience.

Similarly, local knowledge can reside among the groups operating within different communities; hence, the recognition can be extended for the existence of these formal and informal groups involved with the disaster management process. The knowledge and experiences of disaster practitioners remain primarily in the individual domain. As a consequence of the geographical spread of disasters, the experiences, approaches and adopted modalities for disaster management are not codified and remain with individuals as a tacit knowledge (Mohanty *et al.*, 2006).

ISLAND: Inspiring Sri-Lankan reNewal and Development

In recognition of the need for a disaster knowledge networking platform to facilitate interaction and simultaneous dialogue with related expertise *ISLAND* (Inspiring Sri-Lankan reNewal and Development) has been started at the school of the Built Environment, University of Salford, with a 12-month research project funded by RICS. The research is aimed at increasing the effectiveness of disaster management by facilitating the capturing and sharing of appropriate knowledge and good practices in land, property and construction. Due to the broad scope of disaster management-related activities, this initial research focuses on creating a knowledge base on the post-tsunami response, with specific reference to case material in Sri Lanka. The broad aim of the research will be addressed by:

- Creating an infrastructure for developing, sharing and disseminating knowledge about disaster management, particularly mitigation measures, for land, property and construction.
- Developing a network of construction professionals with experience in disaster mitigation practices who can provide advice and support and, if required, on-the-job training and capacity building on a wide range of infrastructure and construction-related issues, such as cost-effective construction methods, waste management, project management, quality assurance, sustainable building materials and alternative water and energy services. The network will promote collaboration between construction professionals and local community-based organisations, thus promoting mutual learning and helping to ensure a people-centred approach to disaster mitigation and reconstruction. Effective collaboration in this manner will facili-

tate consultation to identify and incorporate the needs and aspirations of the affected communities.

- Developing a knowledge base on disaster management strategies arising from post-tsunami recovery efforts, including case materials on post-tsunami response in Sri Lanka. There are many isolated examples of good practice on disaster mitigation and reconstruction, but much of this knowledge remains isolated. The knowledge base will capture good practices and experiences in order to promote mutual learning.

Disaster management is seen as an active, ongoing process of dynamic ventures and needs to be reviewed, modified, updated and tested on a regular basis. There is an urgent need for an organised common platform to capture, organise and collaborate in the dissemination of knowledge on disaster mitigation and to create a versatile, collaborative interface among government, professional bodies, research groups, funding bodies and local communities. The focus needs to be at three levels: institutional (government, ministries, departments, agencies, professional and funding); group (research, voluntarily); and individuals (experts, community leaders). The initial capturing of knowledge relating to disaster mitigation will be done through knowledge-management techniques, including communities of practices, interviews, workshops and a document review. Tacit and explicit knowledge on disaster mitigation will be targeted, but the primary emphasis will be on the tacit component of the knowledge.

In addition to the knowledge base, a dynamic network will be created. The network will use various tools to connect government, institutions, groups and people. In addition, the initiative will involve the development of a web portal to facilitate knowledge collaboration between network members. The portal will provide tools to capture or acquire and organise knowledge, through which the knowledge base will be kept up to date.

The first phase of the research will focus upon identifying an appropriate structure for the proposed knowledge base. Disaster mitigation is recognised as a complex concept with a broad and diverse range of stakeholders. ISLAND is to address the development of an appropriate knowledge base by examining the linkages between different stakeholders and components of disaster mitigation. During the research, a soft system model will be used to break down communication barriers between the stakeholder groups and the associated construction disciplines. The modelling process will include the transcription and verification of stakeholder interview data using NVivo analysis software. Data exploration and interpretation will lead to the development and comparison of fuzzy maps using the software application Inspiration. The subsequent analysis will take advantage of the participative, temporal and qualitative nature of soft systems modelling, which can be used to enhance communication and improve decision making. Fuzzy modelling provides a common language and a shared vision among complex stakeholder groups. It also demonstrates a greater respect for the knowledge and values of local communities affected by disasters, including those local groups from the various construction disciplines. Its domain of applicability is unlimited. The application of soft systems thinking can provide some initial answers to the question of how to improve communication between heterogeneous groups of stakeholders; it can identify capacity gaps in the process of disaster mitigation, recovery and reconstruction. The key to this process is to involve the stakeholders in defining their current and future needs and priorities, and their own proposed solutions. The significance here is the development of more applicable participatory approaches for collecting, analysing and representing information of multiple stakeholders. The focus of soft system modelling in this project is the development of a shared interpretation of a complex problem via personal involvement in modelling. The real power of this approach appears when these relationships are presented in a map format, because then it is relatively easy to see how factors are related and evolving, and how new relationships are formed. Such cooperation is likely to decrease conflicts between stakeholder groups.

Disaster mitigation is a normative goal rather than a rational one based upon measurable predictions or estimates for most locations or contexts. However, if current knowledge and the real concerns of the stakeholders affected are incorporated into the knowledge base structure, then expert views on what the future directions are or should be, are likely to produce sustainable outcomes. Soft system modelling can assist by providing a common framework for different professions to work together effectively.

Conclusion

Disaster management, and construction's specific role in it, are important and evolving disciplines. Knowledge remains fragmented and a lack of effective collaboration and knowledge dissemination is one of the major reasons behind the unsatisfactory performance levels of current disaster management practices. Effective collaboration is required between construction professionals, local government and organisations, and the communities to achieve the common goal of disaster mitigation and reconstruction. This collaboration must ensure that the needs and aspirations of the affected communities are at the centre of these efforts. It is critical that all these groups share a vision and purpose, have clear roles and responsibilities and defined relationships, to achieve the desired outcome.

Defining the vision and desired outcomes will be critical in giving shape and direction to collaboration between these diverse stakeholder groups. It offers the potential for maximising resources, developing sustainable outcomes and greater community ownership and commitment in disaster mitigation and reconstruction. The development of relationships is often viewed as fundamental to the success of collaborations. Expectations must be clear and understood by all stakeholders who are working to protect and rebuild communities. Defining relationships will assist in identifying roles and responsibilities, activities and plans, and, ultimately, reaching desired outcomes.

Future research must aim at increasing the effectiveness of disaster management by facilitating the sharing of appropriate knowledge and good practices in land, property and construction, and building collaboration between stakeholders. Such research should explore the wide range of perspectives from which the construction industry is able to contribute towards improved resilience. It will help to define a more expansive view of the construction lifecycle of infrastructure projects that encompasses the need to anticipate, assess, prevent, prepare for, respond to and recover from disasters. Disaster management needs to enhance knowledge and raise awareness among practitioners and researchers, establish closer linkage between good planning, design, construction, operation, disaster prevention and resilience in communities.

ISLAND represents an early attempt to increase the effectiveness of disaster management by facilitating the capturing and sharing of appropriate knowledge and good practices in land, property and construction. The use of fuzzy modelling will enable the research team to examine a complex problem with diverse stakeholder groups and different construction disciplines. In doing so, it is hoped that the research will facilitate greater collaboration between the construction disciplines and other stakeholders. Although this initial research focuses on creating a knowledge base for disaster mitigation based on the experiences of the Indian Ocean tsunami in Sri Lanka, it is anticipated that the knowledge base will be able to address disaster response and reconstruction in other regions.

References

Banerjee, A. (2005) Tsunami Deaths. *Current Science*, 88(9), 1358.

BBC (2005) http://news.bbc.co.uk/1/hi/world/asia-pacific/4126019.stm, accessed 20 February 2006.

Cabinet Office (2003) *The Capabilities Programme*. http://www.ukresilience.info/contingencies/capabilities.htm, accessed 20 August 2005.

Civil Contingencies Secretariat (2004) *The Civil Contingencies Bill 2004*. The Stationery Office, London.

Coghlan, A. (2004) Recovery Management in Australia: a community-based approach. *Proceedings of the NZ Recovery Symposium*, 12–13 July, Napier, New Zealand, Ministry of Civil Defence and Emergency Management, New Zealand.

Coles, E. (2004) A systems perspective on UK national vulnerability: the policy agenda. *Proceedings of the Second International Conference on Post-disaster reconstruction: Planning for Reconstruction*, 22–23 April, Coventry University.

Davis, I. (2005) What Makes a Disaster. http://tilz.tearfund.org/Publications/Footsteps+11-20/Footsteps+18, accessed on 18 December 2006.

Douglas, M. and Wildavsky, A. (1982) *Risk and Culture*. University of California Press, Berkeley, CA.

Dynes, R. (2003) Finding order in disorder: continuities in the 9–11 response. *International Journal of Mass Emergencies and Disasters*, 21(30), 9–23, Research Committee on Disasters, International Sociological Association.

Fox, A. (2002) Montserrat – a case study in the application of multiple methods to meet a post-disaster housing shortage. *Proceedings of the First International Conference on Post-disaster Reconstruction: Improving post-disaster reconstruction in developing countries*, 23–25 May, University of Montreal, Montreal.

Guzman, M. de (2002) The total disaster risk management approach: an introduction. *Regional Workshop on Networking and Collaboration among NGOs of Asian Countries in Disaster Reduction and Response*, 20–22 February, Kobe, Japan.

Haigh, R. (2003) *Learning on Construction Projects: the Role of Process Capability Assessment*. PhD Thesis, University of Salford, Manchester.

Jigyasu, R. (2002) From Marathwada to Gujarat – emerging challenges in post-earthquake rehabilitation for sustainable eco-development in South Asia. *Proceedings of the First International Conference on Post-disaster Reconstruction: Improving post-disaster reconstruction in developing countries*, 23–25 May, University of Montreal, Montreal.

Jigyasu, R. (2004) Sustainable post-disaster reconstruction through integrated risk management. *Proceedings of the Second International Conference on Post-disaster reconstruction: Planning for Reconstruction*, 22–23 April, Coventry University, Coventry.

Karim, N. (2004) Options for floods and drought preparedness in Bangladesh. *Proceedings of the Second International Conference on Post-disaster reconstruction: Planning for Reconstruction*, 22–23 April, Coventry University, Coventry.

Kendra, J. and Wachtendorf, T. (2003) Creativity in Emergency response to the World Trade Center Disaster. *Beyond September 11th: An Account of Post-Disaster Research. Special Publication No. 39*. Natural Hazards Research and Information Center, University of Colorado, Boulder, Colorado.

Levy, M.P. and Salvadori, M. (1992) *Why Buildings Fall Down*. W.W. Norton and Company, New York.

Lizarralde, G. and Boucher, M. (2004) Learning from post-disaster reconstruction for pre-disaster planning. *Proceedings of the Second International Conference on Post-disaster Reconstruction: Planning for Reconstruction*, 22–23 April, Coventry University, Coventry.

Lloyd-Jones, T. (2006) *Mind the Gap! Post-disaster Reconstruction and the Transition from Humanitarian Relief*. RICS, London.

Mallonee, S., Shariat, S., Stennies, G., Waxweiler, R., Hogan, D. and Jordan, F. (1996) Physical injuries and fatalities resulting from the Oklahoma City bombing. *Journal of the American Medical Association*, 5(276), 382–387.

MCDEM (2004) *Recovery Planning. Information for CDEM Groups*. Ministry of Civil Defence & Emergency Management, New Zealand.

Mohanty, S., Panda, B., Karelia, H. and Issar, R. (2006) *Knowledge Management in Disaster Risk Reduction: The Indian Approach*. Ministry of Home Affairs, India.

National Research Council (1995) *Protecting Buildings from Bomb Damage: Transfer of Blast-Effects Mitigation Technologies from Military to Civilian Applications*. National Academy Press, Washington DC.

Nikhileswarananda, S. (2004) Post disaster reconstruction work in Gujarat on behalf of Ramakrishna Mission. *Proceedings of the Second International Conference on Post-disaster reconstruction: Planning for Reconstruction*, 22–23 April, Coventry University, Coventry.

Nonaka, I. and Takeuchi, H. (1995) *The Knowledge Creating Company: How Japanese Companies Create the Dynamics of Innovation*. Oxford University Press, New York.

Ofori, G. (2002) Developing the construction industry to prevent and respond to disasters. *Proceedings of the First International Conference on Post-disaster Reconstruction: Improving post-disaster reconstruction in developing countries*, 23–25 May 2002, University of Montreal, Montreal.

Pena-Mora, W. (2005) *Collaborative First Response to Disasters Involving Critical Physical Infrastructure*. O'Neal Faculty Scholar Seminar, September 19, University of Illinois.

President's Commission on Critical Infrastructure Protection (1997) *Critical Foundations: Protecting America's Infrastructures*. PCCIP, Washington DC.

Rex, S. (2006) Transforming organisations: from development organisations to disaster response programme: a case study in capacity building. *Proceedings of the First International Conference on Post-disaster Reconstruction: Meeting Stakeholder Interests*. 17–19 May, Florence.

Rotimi, J., Le Masurier, J. and Wilkinson, S. (2006) The regulatory framework for effective post disaster reconstruction in New Zealand. *Proceedings of the First International Conference on Post-disaster Reconstruction: Meeting Stakeholder Interests*. 17–19 May, Florence.

Sevin, E. and Little, R. (1998) Mitigating Terrorist Hazards. *The Bridge*, 28(3), 156–172.

Tobin, A. and Whiteford, C. (2002) Community resilience and volcano hazard: the eruption of Tungurahua and evacuation of the Faldas in Ecuador. *Disaster*, 26(1), 28–48.

UNDP (2005) *The Post-Tsunami Recovery in the Indian Ocean: Lessons Learned, Success, Challenges and Future Action*. Bureau for Crisis Prevention and Recovery, UN, New York.

UNEP (2005) Natural Rapid Environmental Assessment – Sri Lanka. *UNEP Sri Lanka Country Report*, UNEP, New York.

Warfield, C. (2004) *The Disaster Management Cycle*. http://www.gdrc.org/uem/disasters/1-dm_cycle.html, accessed on 22 December 2006.

World Bank (2001) *World Bank and Asian Development Bank Complete Preliminary Gujarat Earthquake Damage Assessment and Recovery Plan.* http://www.worldbank.org/gujarat.

Young, I. (2004) Montserrat: post volcano reconstruction and rehabiliation – a case study. *Proceedings of the Second International Conference on Post-disaster reconstruction: Planning for Reconstruction,* 22–23 April, Coventry University, Coventry.

8 Early design management in architecture

Selecting partners and judging value in design

Leentje Volker

Introduction

Most of the tools and instruments that have been developed to support the design process are based on the assumption that the quality of the product reflects the quality of the process. The fields of construction management, concurrent engineering, lean thinking and new product development elaborate on this assumption by using quite traditional and systemised principles of managing construction projects. These premises fit traditional approaches to construction because the research approaches and research methods used are similar, consequently the importance of the social relationships between the parties is seldom acknowledged (Smyth and Morris, 2007).

This chapter is based on the premise that the best results in design are achieved by collaboration between the design partners and the client. The focus lies on the development of the relationships between the parties involved in the early design phase of a building and the way design values are perceived during this phase. The chapter, therefore, relates to the relationship approach by Pryke and Smyth (2006), and is based on a case study about the selection of architects in a European tendering situation and supporting literature on partner selection and the perception of design quality.

Design management

Within the wide range of design management approaches, a distinction can be made between managing the product, managing the process and managing the organisation (DMI, 1998; Sebastian, 2004). Within the framework of this chapter, design management is project management for the design phase. This means design management starts during the initiation phase with the writing of the brief and decreases during the development of the design. Therefore a design manager could act as a project manager but could also be assigned next to the project manager with a focus on the design only. The aim of design management is to facilitate the realisation of a design which fulfils the expectations of the stakeholders (Prins *et al.*, 2001). Architectural design management is about making existing points of departure explicit, improving upon them through communication and raising awareness. This is done by designing organisational structure, stimulating the creation of a project culture and creativity, developing the brief, drawing up a plan with tasks and responsibilities and finishing the design phase (Doorn, 2004). The most important actors in this initial phase of the construction process are the client, the architect and their advisors from different fields (e.g. interior design, construction, building costs, briefing and project management). The more complex the assignment is and the more people and organisations are involved, the more need there exists for design management (Doorn, 2004). When production processes are getting more standardised and the influence on the design reduces, design management is being replaced by project management for construction. Methods such as PRINCE2 and PMBOK® all express the need for a structured overview of the management aspects of cost, time, quality, information and communication. These aspects seem to be inconsistent with the rather intuitive and chaotic design process and the way professional service firms act and interact. Within the range of management activities during the design phase a distinction can be made between the 'hard or tangible' steering activities – such as the use of planning techniques and quality management – and the more 'soft or intangible' aspects of steering, which include leadership, design vision, social relationships and team bonding (Volker and Prins, 2006).

In project management particular attention is given to the tangible aspects. However, over the last few years the focus on the more intangible aspects and elusive phases of project management has increased. This could be partly due to the still high costs of failure while the systems and methods keep improving. Another reason could be the growing insights into the similarities between managing and designing, and the importance of involving all stakeholders in both processes. For example, Sebastian (2005) acknowledges that design and management are both knowledge-intensive social activities, which work with and within uncertain situations to deliberately initiate and devise a creative process for shaping a more desirable reality. Both management and design activities require analytical, synthetic and evolutional thinking processes. And both activities are necessary to reach the project goals.

In the Netherlands the principle of process management is gaining appreciation. This view on project management focuses on the guidance of the interaction processes between equal stakeholders with unclear and diverse goals yet with the intention to raise the quality of the current state (Bruijn *et al.*, 2002). Collaboration, communication, influence and persuasion are important aspects (Bekkering *et al.*, 2001). Process management is mostly applied in city area redevelopment projects such as railway stations and transformation projects. In the case of design management, process management techniques can be applied in the early stages of the building project to find ways of collaborative working and involving stakeholders in the decision making and development of the brief. For example, for a city to build a new theatre on a sensitive location in the city centre, user participation and intensive communication with the external stakeholders could decrease the opposition and increase the support among the citizens. Process management could facilitate the implementation of the design in this manner.

Whilst the client is one of the most important stakeholders in the design process, most clients are inexperienced with designing and constructing a building. Therefore they often hire an advisor to assist, who is frequently functioning as the project manager too. Krenk (2006) mentions that a medium level of structure and a high degree of collaboration between the project owner (client) and the project manager (advisor) are key factors for the best project performance. Realising the importance of cooperation, coordination and clear communication (Emmitt, 1999; Walker, 1998), in Denmark they are experimenting with an additional client consultant (Krenk, 2006). The client consultant functions as the single channel of communication giving the client advantages in regard to controlling the information flow, getting sufficient professional assistance and making it possible for the client to decide on the degree of day-to-day involvement. Yet it adds a link to the chain of communication and increases the complexity of the project organisation (Krenk, 2006). Preliminary results from Krenk's case study show that this extra link next to the project manager is not seen as problematic but moreover as beneficial to the level of performance. Clarifications of the roles and responsibilities have ensured mutual understanding between the client and the project manager. The primary concern of the client representative was therefore to ensure the client's interests: satisfactory quality within the programme and budget. The issues of financial empowerment, conflicting interests and responsibilities are interesting topics for further investigation.

According to Gray *et al.* (1994), an essential condition for effective design management is knowledge and understanding of architectural value and the essentials of design processes, designers and design organisations. Collaboration is a far richer process than teamwork (Gray and Hughes, 2001). Collaboration requires people to work together freely to maximise their potential. This can only happen when there is mutual trust and respect for each other's capabilities and a long-term relationship with the designers. It must allow the continual exchange of information and knowledge without any barriers being put in the way.

Collaboration enables adjustment to the roles of the client, manager and architect for design development. According to the 'wheel of dominance' (Gray and Hughes, 2001) the client (or its representative) should take the lead in developing the brief. The architect then takes on the leading role in developing the design sketch and concept of the design. In the detailed design and construction phase the project manager should be in a leading position. This concept is based on two levels of responsibility: the associated authority for decision making, and responsibility for the interface with other organisations. The task of the design manager is to ensure that the organisation of the design process is structured appropriately and that there are sufficient integrating and coordinating mechanisms for meaningful progress. This implies a design manager is working next to the architect, project manager and the client. This design manager could also act as an external chief decision maker to integrate and manage knowledge held by the stakeholder groups (Kestle and London, 2002).

Selecting team members

This project team of a design project usually consists of the client (or its representative), a project manager and an architect. The client has to set up the project and put together a project team including a project manager and a designer at least to execute the project. Based on a literature study Chan *et al.* (2004) found that the characteristics of the client, the composition of the project team, and the level of experience of the partners are human-related factors that are of influence on the performance of the project. In assembling a team, careful consideration should be given to the level of professional experience, the design experience and the personalities of the team members, and to whether the team is sufficiently multi-disciplined (Shen and Liu, 2003). Walker (1998) suggests that selection of a client's representative should be based on the capacity of the client's representative to gain the confidence of the project team. According to Jha and Iyer (2006) the most important traits that should be looked into when selecting people as coordinator of a project are human relationship and team-building skills. From a phenomenographical perspective Chen and Partington (2006) identify three basic cumulative and hierarchical conceptions of project management work in the UK construction industry:

- Project manager as planner and controller.
- Project manager as organiser and coordinator.
- Project manager as predictor and manager of potential problems.

In the same study Chen and Partington (2006) demonstrate how the project managers' way of experiencing their work, constitutes their competence at work by taking the project management work and the worker as one entity. They state that the intangible aspects of the person rather the conception of their work seem important in project management. Based on the fact that the initial design phase is full of uncertainties and potential conflicts between stakeholders, the project manager selected by the client in the early design phase should rather be able to organise, predict and act as a manager of potential problems than being a planner or controller.

Selecting an architect for relatively large building projects requires a different procedure than the selection of a project manager. According to the European Regulation 2004/18/EG every semi-public organisation in the European Union which wants to select an architect for over 211 000 Euros as at July 2007 (147 000 Euros for central governmental bodies) has to use a European tendering procedure. These regulations are developed to stimulate European integration. Basic principles during this procedure are public, transparent and non-discriminatory. Several procedures are possible: the presentation of a design vision, the presentation of a design idea, the presentation of a sketch design or the presentation of an almost detailed design. Depending on the procedure and willingness of the client, the architects receive a financial compensation for their activities. During these procurement procedures the client, usually assisted by the project manager and/or an additional consultant, selects and employs architects, and then commissions them to design the facility. Sometimes the architects have a say in introducing the rest of the advising professionals. A direct relationship between the client and the architect is seen as an essential element to achieve quality in design (Lewis, 1998). Having a clear well-understood objective is crucial for directing the process and concentrating participants' efforts (Swink, 2003). Ling (2003) found that, according to project managers, architects should have knowledge of economical designs and constructability, producing designs which have functional quality, gaining adequate job experience, and producing design drawings and obtaining statutory approvals speedily. Project managers would also like architects to accept them as project team leaders, follow and respond quickly to their instructions and be loyal to them. Architects attached lesser significance to many of the attributes stated above than the project managers did (Ling, 2003). So there seems to be a discrepancy between the architects themselves and the project managers that usually select them on behalf of the client in relation to the attributes needed to be a professional architect.

The design of construction projects is a collective effort involving a team of specialists from different organisations (Cheung *et al.*, 2001). Design teams usually consist of the architect, a structural engineer, a building services engineer and a quantity surveyor. Generally in the

Netherlands the architect acts as the leader of the design team. This leadership should not be confused with the project manager of the project team. The leader of the design team does not always act as a member of the project team. The leadership of the design team could affect the productivity of the design team and therefore the project success. In an empirical survey Cheung *et al.* (2001) found that in Hong Kong the 'charismatic' and 'participative' leadership behaviours of the design team leader primarily determine the satisfaction of the team members and the way in which they feel proud to be affiliated with the team. Leadership plays a significant role in synergising diversified inputs in the design integration process. Hence the potential tension between the project manager, the leader of the design team and the paying client. Most of the members of the design team are contracted by the client and therefore not hierarchically linked with the design team leader. A lack of direct contractual relationships makes the line of authority subtle. The highly interrelated design task can only be achieved through concerted effort by a motivated design team. The results point to the uniqueness of project design work (Cheung *et al.*, 2001). Furthermore it should be noted that in most public organisations 'the' client does not exist. Public clients usually put together a steering committee to provide input in the project team. This steering committee usually represents several user groups and other stakeholders.

Architectural quality and value

The final goal of collaboration in architecture is to construct a product that fulfils the wishes for the build quality articulated by the stakeholders (Prins *et al.*, 2001). Most of the latest literature on build quality originates from the UK and is partly related to the discussion about the design quality indicator (DQI) (e.g. Whyte and Gann, 2003; Slaughter, 2004). Quality in building design embraces all of the aspects by which a building is judged, such as uniqueness, functionality or durability. The triangle of Vitruvius' 'utilitas, venustas and firmitas' – usability/functionality, beauty/authenticity and construction/durability – has been a source of inspiration, and has been elaborated upon in several publications (e.g. Duerk, 1993; Voordt and van Wegen, 2005). Prasad (2004) describes quality as the achievement of a totality that is more than the sum of its parts. He argues that truly excellent design quality can only be achieved when the three quality fields of functionality, build quality and impact all work together (Figure 8.1). Added value is created by the overlap of two of the fields. From a value management perspective, value depends on balancing the three factors of time, cost and quality against client requirements, whilst retaining the basic ideal, that is, to

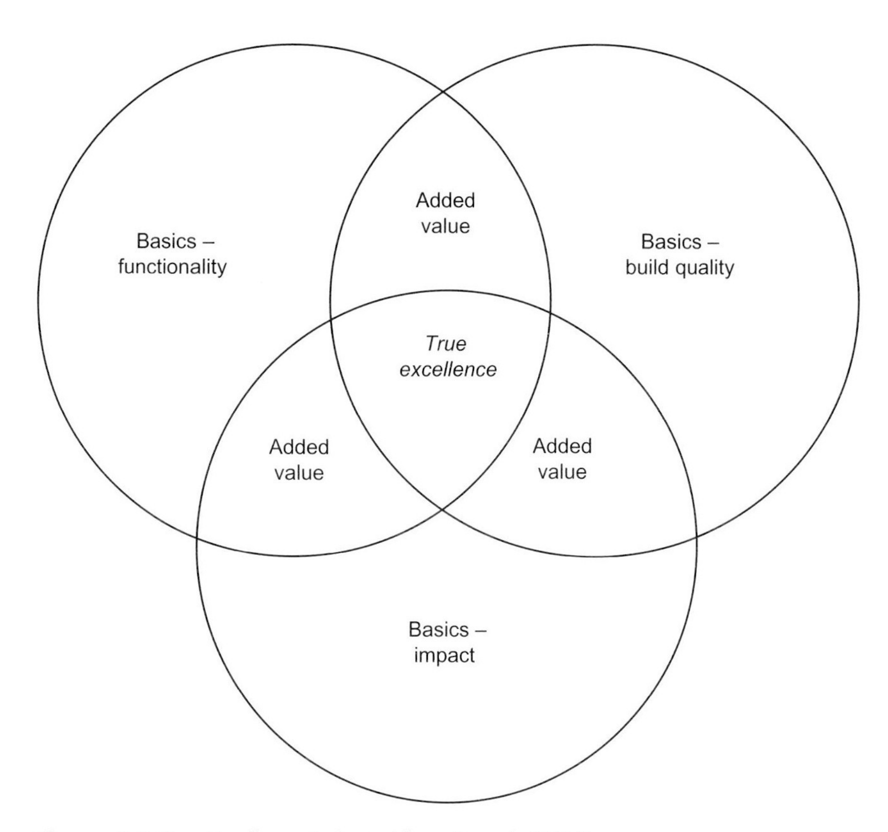

Figure 8.1 True Excellence (adapted from Prasad, 2004)

complete the project at minimum cost, in the shortest possible time and to the highest possible standard. Value always involves a relative and balanced consideration of tangible and intangible costs and benefits, as well as a willingness to give up in order to gain (Best and deValence, 1999).

From the view of 'quality is in the eye of the beholder' it is mainly customer satisfaction that determines build quality and provides for the intended values. The concepts of Kano provide insight into the customer requirements (Walden *et al.*, 1993). The stated quality of the product is one-dimensional, combining functionality and satisfaction. Basic requirements must be seen as the quality of the product, while attractive and satisfying products include something of a surprise and therefore could create love for the product (Figure 8.2). Satisfying products would involve something new, exciting and surprising, something to be loved. This seems to be something which can only be perceived and judged by the client himself holistically and intuitively, because the structured requirements which can be managed and measured are exceeded.

Figure 8.2 Customer defined quality (adapted from Walden *et al.*, 1993)

Hence it would be interesting to know whether it would be possible to steer this creation of love for a building. Managers as well as designers need to be more engaged in the delivery of outcomes, whilst the willingness to increase short-term costs for long-term gain has to grow (Macmillan, 2005). The case study as described in this chapter is part of the explorative phase of an ongoing research project on the added value of design management in highly ambitious and complex architectural design projects. Instrumental and human management activities, design leadership, decision making, competencies of the actors and the perception of value creation are key aspects in this project (Volker and Prins, 2006).

Research method and case description

To provide some more insight into the relationship between collaborative activities during the first design phases and the value creation of the design, a single instrumental case study has been performed (Stake, 1995). The case study is an in-depth study of the phenomenon of a

European design tender. Nine semi-structured focused interviews (with the client, the architects and the project manager), field notes of three meetings and observations during ten presentations were used to collect empirical data. The case study was taken from a realistic view which makes it possible to value perceptions of the actors about the procedure and decision-making processes that took place (Smyth and Morris, 2007). From the view that the perception of the product and process determines the success of a project and the creation of value, there are too many personal factors and too much uncertainty that act upon these situations to adopt a more structured and deterministic paradigm.

The aim of this case study is to understand the process of judging value, applying management in design, and to generalise across a larger set of units (Gerring, 2004). During this phenomenon the design parties and the client are the units of study. This research method answers to the need of exploration of the complex relationship based on causal mechanisms between design processes and design values in the context of human perception. One of the consequences of using a case study method could be the relatively limited generalisability of the results. However the case study was set up according to the principles of Yin (1984), Flyvbjerg (2004) and Stake (1995) for validity and reliability as much as possible. A pilot study was conducted on a European tender for an Institute for the Blind in Zeist, the Netherlands. During this pilot study an observational method was developed to list the activities of the designers and make notes of the meetings. Also an interview protocol was developed and tested. The results of the case study are used in this paper partially to interpret interview data from the actual case study.

This single case study is about a European tendering procedure to select an architect for the design of the new town hall and library for the historical city centre of Deventer. Deventer is a city in the eastern part of the Netherlands near the river Ijssel and has almost 100 000 citizens. The design task called for 18 300 m^2 of new space and was challenging because of the central location and small streets of historical character. Decisions needed to be made about the retention or demolition of existing office and theatre buildings, and the City of Deventer wanted to utilise the professionalism of the architects by giving the architects as much freedom as possible in providing a design solution. The building has to be finished by 2010. The City of Deventer is to be the end-user of the building. In this case study the steering committee of the City of Deventer is referred to as the client.

The case was selected because of the complexity of the design task and the public availability of information. The design task could be labelled as challenging because of the historical setting and volume of the building. Therefore a lot of nationally renowned design firms joined the tender. The need for design management increases as the complexity of the design and the number of organisations involved expand (Doorn, 2004). In this case a lot of stakeholders were directly involved because of the public character of the project; citizens, politicians, employees, expert groups and other external stakeholders were also involved. Because of this, the use of design management is expected. An advantage of the European tendering procedure is the obligation to be transparent, meaning documents are accessible and reported publicly. During the pilot study we found that sensitive information was hard to find and the client was reluctant to participate in the wider agenda. Yet the City of Deventer as the end-user was willing to participate in this case study because it fitted with their objective to be transparent. This same advantage makes this process relatively sensitive for political issues.

Selecting an architect, following the European tendering procedure, was achieved as follows:

- The client expressed the need by publishing an announcement on the Internet and through official channels. This announcement included a selection manual with the selection criteria and requirements of the firms the client wanted to join the tender.
- Interested parties could register interest in the tender by providing the requested information.
- The client selected a number of companies, in this case five, and invited these architectural firms to make a sketch design based on the selection manual.
- Designers submitted their scale model and additional descriptions of the design. In the attachment a cost estimation had to be included, as well as a plan of action for further collaboration.
- The client decided on the design that best fitted their needs behind closed doors. For this special purpose the client appointed a selection committee. This selection committee consisted of representatives of the democratically chosen parties in the City Council. They had advice of several users and expert groups at their disposal. Consequently the final decision was based on a judgement from the selection criteria based on information provided by the documents and scale model, as well as presentations by the designers themselves.
- The design firm securing the job officially can be considered as the competition winner.

The final aim of this case study is to evaluate the judgements in the context of the design processes that have taken place and compare the selection criteria in the selection manual to the final outcome of the tender. This process is worked out in Figure 8.3. Interviews were used to reconstruct the design processes in the event of being unable to more immediate access to 'real time' data.

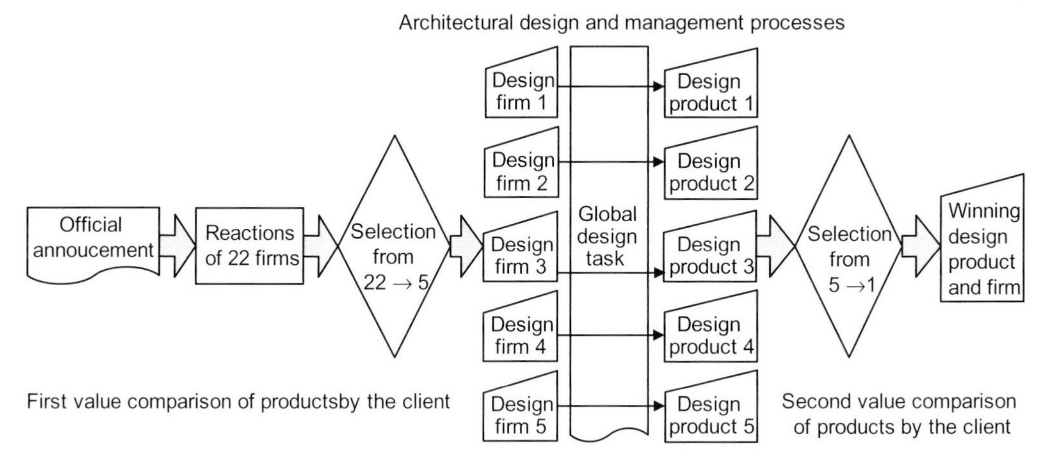

Figure 8.3 Partner selection process in European tendering procedure of the City of Deventer

Results

This case study produces results on several aspects of a design project. The judgment on values of architectural design is one of the most important aspects. Because this case concerns the context of tendering, a lot of results refer to the decision making about the selection of parties and the applied European procedures. The description of the results will focus on the value judgment and the managerial activities of the architects as well as the activities of the client enabling the selection process.

For Dutch firms a European tender provides an infrequent or unique chance to design a city hall, therefore these kinds of competition attract a range of practices. The architectural firms that were participating in this tendering procedure were all stimulated by the complexity of the design task. They stated that they were attracted by the great building volume in a historical city centre, the ambition of the City of Deventer to combine the city hall with a library and the mixture of old and new. With all participants the final decision to join the tender was made between the partners of the design firms. Just after the firms were informed about being selected as the final five competitors, the tender was officially assigned to one of the partners. Although the partner carries the main responsibilities, he is assisted by a design team including a project architect. This project architect can be seen as the internal project coordinator and one of the main designers. Only one of the firms mentioned the start of the design by making a project plan with the proposal of a team structure, planning and budget. According to the interviewed architects, the design team was put together based on availability of the designers and characteristics of the design. This means that some firms had to hire external parties to provide for a three-dimensional design, the scale model or descriptive texts. Some firms had all the required capacities within the organisation.

In most of the firms the first sketches for the Deventer site were made by the partners of the design firm. Sometimes the project architect and partner discussed the design options but the partner had the final say in the direction of the design. As the process progressed, the role of the partner changed from idea generator into guiding the design vision and advising the project architect on general and specific briefing requirements. The amount of time spent on the actual design by the partner seemed to decrease during the process while the other designers spent increasing time on working out the design. The project architect was primarily looking after the precise briefing requirements and available information. All presentations of the sketch designs and design visions to the client were done by the partner themselves, sometimes assisted by the project architect. One firm completed the presentation with a vision from the bureau manager on the project management aspects. During this presentation, emphasis was given to the vision and collaboration between the architect and the client during the design process. Other firms paid relatively less attention to this aspect of the assignment.

According to all the architects, the actual design started with an analysis of the design task. One of the architects described the development of the design for Deventer as a combined interpretation of the brief and the site. His design team got the information they needed from the brief and a very thorough analysis of the historical site (because they could not interact with the client directly as usually is the case). Because in this case the same site was being used by all the firms, the interpretation of the site and the brief seems to define the actual character of the architects' work and the differences between them. During the interpretation of the design task, all architects missed the input of the client. In reflecting interviews they all stressed the fact that they were not sure what their client was looking for. Yet the project manager and the client emphasised in their reflective interviews that they were definitely looking for an icon. According to the client this ambition had been widely announced during the kick-off meeting with all the participants of the tender competition. These perceptual differences were probably caused by the fact that there was no interaction possible between the client and the architect during this uncertain and probing phase in design.

In the case of the City of Deventer, the selection of an architect for their city hall was conducted in two rounds with different procedures. In the first round, a selection had to be made out of 22 parties who had shown interest and officially answered to the requirements. The client systematically reviewed the firms, asking the question, 'Do we know this firm and do we like their work?' They used the selection manual as a helping hand. Hardly any discussion took place about the first three selected firms. This is contrary to the final two firms, which were selected out of the lesser ten firms that the client felt suitable for the second round. The empirical results do not show a direct connection between the decisions made in this first-round selection and the winner of the second round. It could be that these favourite firms might have had a better chance of winning because all members of the client committee were in favour of them. The final voting decision in the second round was made by a selection committee of representatives of all political parties of the City Council. During the second round other stakeholders, such as the employees of the city, the library, the experts committee and the city planners, were involved in decision making by providing advice to the selection committee. Almost all advising groups agreed on the high design quality of all five designs. Not all advising groups were unanimous in favour of one of the designs but most of them included the design of the final winning firm, Neutelings Riedijk. During one of the client interviews the process of expert judgment was described as first judging the building upon aesthetics, especially the exterior, then functionality, and finally the way they felt the design had special attributes.

Because the City of Deventer wanted a transparent process, the members of the selection committee justified the opinion of their party about the design during an open debate the evening before the final voting decision:

1. The *aesthetics* were judged in terms of being striking, surprising, original and daring; the allure of the design, the insertion and rejection in the environment and the recognisability of the building were key factors, adding value to the image of the city as a design that 'touches the heart'.
2. *Functionality* of the design mostly came across in reflection on the size of the building, the readability and orientation within the building, the amount of daylight and transparency, the flexibility and possibilities of separation of the interior and the materials used. Functionality issues mentioned were the location of the library in terms of how the designs 'felt' and fitted into the historical environment, the relation between light and space, the fitness to use an innovative office concept and the division of the space. There were differences between parties. The management and staff representing the library and the municipality had distinct views based upon functionality and the user representatives and citizens also had different views as to what constituted a good design.
3. The judgment of the *politicians*, as representatives of the political parties of the city council, was partially based on their personal experiences during the presentations of the designs and partially

based on the advice of users and expert groups. They vigorously debated the designs from their respective viewpoints. During the debate they felt little responsibility to take account of the advice of the stakeholders in their decision because of the large differences between the stakeholders.

4. All designs were *counterweighted* and approved by an external financial party and the urban planning department of Deventer. The financial feasibility of the designs had hardly been mentioned during the public debate. Some parties did stress the limited flexibility of the budget.

The winning design of Neutelings Riedijk was described in terms of love, heart, feeling, surprise, and being exciting, beautiful and adding value to the city. The official press release of the City of Deventer expressed the design of Neutelings Riedijk as a most appealing design, fully functional and full of contrast. According to the City of Deventer the design fitted seamlessly into the historical context of the city, using design solutions that were sometimes surprising as well as excellent. Most of the architects were somewhat surprised at the final judgment. In the reflective interviews the client as well as the project manager and most of the architects revealed some critical notes about the functionality and the potential costs of the winning design. These critical notes seemed to have been repressed during the political decision-making process. In this case the feeling about the aesthetics seems to have taken over the rationale of the budget and functionality.

How effective were the collaborative relationships in this case study? The composition of the advising client committees differed in levels of professionalism and conflicting interests. For some members of the client committee it proved hard to evaluate the sketch design to understand the vision of the architect. During the presentations, members focused at times on very human issues of detailed functional design, such as the location of the bicycle shed and staff entrance, while the sketch design did not attempt to address these kinds of decisions. Selection criteria mentioned in the selection manual only cover the basic requirements to be judged by the client. The reflective interviews with the client and the project manager afterwards seemed to affirm that, especially during the second round, not all of the selection criteria were used explicitly in the judgment. Because the client could not express some of their criteria in objective terms, they referred to the criteria of 'most appealing design' as their main decision criteria. And the 'most appealing design' seems to include 'intangible' criteria, such as personal connection, faith in and affinity with the architectural firm, design and designers. The client confirms this finding by expressing that the design product was intertwined with the designer and the design firm. Therefore the winner reflects the design firm which the client believes will best fulfil their needs and provide them with the right architectural value.

The architects explained the decision making of the client by using the terms 'trust' and 'believe in my blue eyes' when talking about the selection of partners. They accepted the fact that the client had 'fallen in love' with the design because they believe it's the feeling of the client that should decide on the best design. For architects the problem with these kinds of procedures is that they cannot estimate their chances of winning a tender. That is why they use the terms 'ritual dance' and 'seduction game' in these matters. True excellence seems not to be about answering to all functional, durable and aesthetic requirements. In this case study the client concentrated on the design, especially the aesthetics of the exterior, and in choosing a design took on board an architect as the design partner.

Conclusion

This first explorative case study provides a greater understanding about the aspects of management and quality perception during the initial phase of design and the development and judgment of the value of a design product. Although the ability to generalise from this case study might be limited, it contributes to the understanding of existing tensions and difficulties in the selection of partners and collaboration in design and construction. Data has been gathered by combining several research methods. Even more methods were used than were originally thought during the design of this first explorative step. This way of combining methods, also called triangulation, seems to complete the picture of the design and judgment processes reasonably well. It therefore contributed to the validity and reliability of the study. It also contributed to the learning process of the researcher and other actors in the case study. The architects as well as the client were very interested in the results and insights from 'the opposite side' of the case.

Hardly any traditional management activities were identified during this phase of design. Architectural firms do not seem to work with official quality and management systems like ISO but every action seems to be based on assumptions that are embedded in the culture of the firm. These cultures seem to differ by firm and reflect their style and qualities. One of the architects stated 'architecture is a people business'. In selecting a project partner, intangible qualities seem to influence decision making greatly. The results of the first step support the idea of looking more closely at the management principles used in professional service firms and the management of project firms. It can be concluded that design management activities were located in a broader set of social and political conditions that engaged a range of important stakeholders, which together constituted a management of projects activity. In creating architectural value, communication, trust, professionalism and potential leadership seem important.

Supposing that transparent, public and non-discriminating implies objective application of the criteria, this case study shows that applying selection criteria in tendering situations is far more complex in practice than is portrayed in the official tendering regulations. Applying the selection criteria in a subjective manner seems to simplify the decisions greatly.

For the City of Deventer, the most valuable design did not exactly match the preconceived selection manual, but did reflect opinion about their future partner. The City of Deventer found an architect to design their new city hall annex library in a politically robust way. They hired a project manager to facilitate the tendering procedure, make the necessary decisions and run the process. The criteria the client used in their judgment had a lot in common with the attributes that are mentioned in Kano's work on client satisfaction (see Walden, 1993 and Figure 8.2). Even though all designs answered the basic requirements, the winner of this contest was chosen because of the surprising design solutions that created a feeling of love for the product. Because of the status of the design it is hard to decide on the actual fit of the design concept with the final brief of the City of Deventer. The future will tell if the tension between the budget, brief and design will produce a beneficial outcome.

For future collaboration, the people side of the project needs to be explored and utilised between the actors. According to all interviewees respect, openness, responsibilities and trust seem more important in creating value in design than planning and costs. Of course boundaries need to be set by the client and professionalism needs to be shown by the architect and the project manager, but creating an open, ambitious but critical dialogue between the client and the designers seems to be the basis for good design. The project manager seems to have an important task here. He or she should represent the client's needs but empathise with the designers. He or she has to integrate the wishes, needs and available means of the client with the solutions and ambitions of the design team. These kinds of competencies require insights in architectural design but above all excellent leadership, social skills and the will to collaborate.

Acknowledgement

The author would like to thank all participants, especially the City of Deventer, for their contribution to this case and Dr. Kristina Lauche for her assistance during the preparation and analysis of the data.

References

Bekkering, T., Glas, H., Klaassen, D. and Walter, J. (2001) *Management van Processen – Succesvol Realiseren van Complexe Initiatieven (Management of processes).* Het Spectrum, Utrecht.

Best, R. and deValence, G. (1999) *Building in Value – Pre-Design Values.* Arnold, London.

Bruijn, H. de, Heuvelhof, E. ten and in't Veld, R. (2002) *Process Management.* Academic Service, Schoonhoven.

Chan, A.P.C., Scott, D. and Chan, A.P.L. (2004) Factors affecting the success of a construction project. *Journal of Construction Engineering and Management,* 130(1),153–155.

Chen, P. and Partington, D. (2006) Three levels of construction project management work. *International Journal of Project Management,* 24(5), 412–421.

Cheung, S. O., Thomas Ng, S., Lam, K.C. and Yue, W.M. (2001) A satisfying leadership behaviour model for design consultants. *International Journal of Project Management,* 19(7), 421–429.

DMI (1998) 18 views on the definition of design management. *Design Management Journal,* 9(3), 14–19.

Doorn, A. van (2004) *Ontwerp/proces (Design/Process).* Uitgeverij SUN, Amsterdam.

Duerk, D.P. (1993) *Architectural Programming: Information Management for Design.* Van Nostrand Reinhold, New York.

Emmitt, S. (1999) Architectural management – an evolving field. *Engineering Construction and Architectural Management,* 6(2), 188–196.

Flyvbjerg, B. (2004) Five misunderstandings about case-study research. In: Seale, C., Jaber G.G., Gubrium, F. and Silverman, D. (eds.) *Qualitative Research Practice.* Sage, London.

Gann, D.M., Salter, A.J. and Whyte, J.K. (2003) Design quality indicator as a tool for thinking. *Building Research & Information,* 31(5), 318–333.

Gerring, J. (2004) What is a case study and what is it good for? *American Political Science Review,* 98(2), 341–354.

Gray, C. and Hughes, W. (2001) *Building Design Management.* Butterworth-Heinemann, Oxford.

Gray, C., Hughes, W. and Bennett, J. (1994) *The Successful Management of Design.* Centre for Strategic Studies, University of Reading.

Jha, K.N. and Iyer, K.C. (2006) What attributes should a project coordinator possess? *Construction Management and Economics,* 24(9), 977–988.

Kestle, L. and London, K. (2002) *Towards the Development of a Conceptual Design Management Model for Remote Sites. IC+GLC-10.* Gramado, Brasil.

Krenk, C. (2006) The communicational aspects of the building process – a necessary expansion of the scope. In: Scheublin, F.W.J. and Pronk, A. (eds.) *Adaptables 2006 – International Conference on Adaptable Building Structures.* TU Eindhoven, Eindhoven.

Lewis, R.K. (1998) *Architect? A Candid Guide to the Profession.* MIT Press, Cambridge, Mass.

Ling, Y.Y. (2003) A conceptual model for selection of architect by project managers in Singapore. *International Journal of Project Management*, 21(2), 135–144.

Macmillan, S. (2005) *Better Designed Building: Improving the Valuation of Intangibles*. Eclipse Research Consultants.

Prasad, S. (2004) Clarifying intentions: the design quality indicator. *Building Research & Information*, 32(6), 548–551.

Prins, M., Heintz, J.L. and Vercouteren, J. (2001) Design and management: on the management of value in architectural design. In Gray, C. and Prins, M. (eds.) *Value through Design*. CIB, Reading.

Pryke, S.D. and Smyth, H.J. (eds.) (2006) *The Management of Projects: a Relationship Approach*. Blackwell Publishing, Oxford.

Sebastian, R. (2004) Critical appraisal of design management in architecture. *Journal of Construction Research*, 5(2), 255–266.

Sebastian, R. (2005) The interface between design and management. *Design Issues*, 21(1), 81–93.

Shen, Q. and Liu, G. (2003) Critical success factors for value management studies in construction. *Journal of Construction Engineering and Management*, 129(5), 485–491.

Slaughter, E.S. (2004) DQI: the dynamics of design values and assessment. *Building Research & Information*, 32(3), 245–246.

Smyth, H.J. and Morris, P.W.G. (2007) An epistemological evaluation of research into projects and their management: methodological issues. *International Journal of Project Management*, 25(4), 423–436.

Stake, R. (1995) *The Art of Case Research*. Sage Publications, Thousand Oaks.

Swink, M. (2003) Completing projects on-time: how project acceleration affects new product development. *Journal of Engineering and Technology Management*, 20(3), 319–344.

Volker, L. and Prins, M. (2006) Linking design management to value perception in architectural building design. In: Hamblett, M. (ed.) *International Built & Human Environment Research Week – 6th International Postgraduate Research Conference*. Delft.

Voordt, D.J.M. van der and Wegen, H.B.R. van (2005) *Architecture in Use: an Introduction to the Programming, Design and Evaluation of Buildings*. Architectural Press, Oxford.

Walden, D., Berger, C., Blauth, R. and Boger, D. (1993) Kano's methods for understanding customer-defined quality. *Center for Quality of Management Journal*, 4(2), 1–37.

Walker, D.H.T. (1998) The contribution of the client's representative to creation and maintenance of good project inter-team relationships. *Engineering Construction and Architectural Management*, 5(1), 51–57.

Whyte, J.K. and Gann, D.M. (2003) Design quality indicators: work in progress. *Building Research & Information*, 31(5), 387–398.

Yin, R.K. (1984) *Case Study Research: Design and Methods*. Sage Publications, Beverly Hills.

9 Civil construction supply chain management policy to support collaborative relationships in public sector procurement

Kerry London and Jessica Chen

Introduction

Achieving performance improvement in the construction industry through the improved performance of the supply chains has been subject to considerable international debate. This normative approach to improve firm behaviour and ultimately industry performance through the development of supply chain clusters or integrated supply chains has been discussed in many public sector policy documents and in the academic research community since the late 1990s (London, 2004). However, it has been difficult to see any real examples where this concept has had any major impact, or where the improvements have been measured and/or monitored. This chapter seeks to identify the problems arising from implementation of government strategic purchasing policies for supply chain management.

Governments generally understand their role as a key purchaser, and the impact that they can have on firm behaviour in an industry in relation to projects, but perhaps do not exercise this role to the full extent. Government project procurement policies and strategies with sound principles have been developed. However, one of the greatest difficulties is the lack of effective implementation of strategic purchasing policies to manage change and improve performance and productivity across entire sectors within the property and construction industry. Yet purchasing policies are typically developed for the whole of government, that is, generic policies for all the agencies. Many of these policies focus particularly on first-tier supplier management, which is a very useful first step. However, there is still a lack of understanding of the interdependency between firms in the supply chain and the role that the interrelationships between firms have on the overall performance of the industry. Effective implementation requires deep structural and behavioural change amongst large groups of firms, and not simply the first tier of suppliers who supply to the government. Civil construction infrastructure agencies typically focus on this first tier as they relate to project procurement strategies and tend to leave subsequent tiers to contractors and consultants and the market forces, that is a non-market interventionist approach. The dichotomy of complete market intervention and non-market intervention is, however, a simplistic approach to project procurement.

With the necessary tools to evaluate and monitor supply chain performance throughout the sector, a direct contract with consultants and contractors can provide the framework to manage the remainder of the supply chain through a range of incentives and rewards (Kenley *et al.*, 2000; Pryke and Pearson, 2006). Corporations manage the supply chain through a range of relationship types with varying degrees of rewards and incentives. Typically they develop a clear understanding of their risk levels related to their level of expenditure for different options. Government civil infrastructure agencies have yet to understand the full impact of their purchasing power across entire portfolios of projects, realise their risk versus expenditure profile in specific sectors and develop suitable relationship types to their risk versus expenditure profile. Effective supply chain management by large private sector corporations involves assessment of different levels of risk and expenditure and the development of a range of relationship types.

Governments can play a significant role in changing the structural and behavioural characteristics of industries through the use of the supply chain concept, even if to date it has had limited understanding, support or implementation. The relationship between government policies affects the character of the industry (Cox and Townsend, 1998) in three main ways. Policy instruments (direct intervention and regulations or indirect development and monitoring of policy) can encourage competition or alternatively restrict practices in certain areas (Warren, 1993). Second, government as a large client of civil infrastructure projects, can impact upon the state of the industry by virtue of their overall demand patterns through the allocation of annual budgets and capital works programmes. Finally, as a large client their purchasing power can be a mechanism to induce certain types of firm behaviour through contracts on individual projects. It is proposed that governments should drive the agenda because of their broad overarching role to ensure improved industry performance from a holistic perspective, rather than market suppliers who typically take an individualistic approach and do not take the initiative nor should be expected to improve all parts of the supply chain, unless rewarded to do so. Public sector clients, like their large corporation counterparts, can develop management strategies in relation to their supplier groups but, of course, public sector organisations have different influences affecting how this is achieved. Developing a supplier group strategy map (London, 2007) relies upon firstly understanding the economic characteristics of the supply and demand markets followed by developing strategies to achieve productivity and innovation performance improvements.

Therefore, the general research question this chapter addresses is *how can public sector clients develop supplier group strategy maps?*

This raises three key questions to consider in relation to supply chain performance and industry productivity specifically for public sector organisations:

1. How do we improve sector productivity and innovation performance given existing structural and behavioural characteristics in an industry with a large number of interdependent firms?
2. What role can the government play in improving sectoral performance as both a large client and a regulator or policy maker?
3. What are the difficulties related to public sector supply chain and strategic procurement policy development and implementation?

The empirical study for this chapter involved two detailed case studies of supply chains clustered around two major industry players: a state government civil infrastructure agency and the largest local government civil infrastructure agency in Australia. The study also involved a third agency: the agency responsible for whole of government procurement and building policy and, separately, polices relating to the procurement, management and maintenance of government buildings. There is a 'Procurement and Building Agency' (The Department of Public Works) but it does not have responsibility for whole of government construction policy beyond that applying to government buildings. Within that Agency there is a Building Policy Unit, and, separately, a Procurement Policy Unit. This chapter discusses partial results of the state government civil infrastructure agency case study. The case study involved the interactions between the various players in the supply chain for the pre-cast concrete sector, the civil construction industry policy development unit and the civil infrastructure agency while exploring the three key highlighted issues. This chapter is limited to the civil construction government agency and their interactions and/or influence on the pre-cast sector.

The key concerns of the state government civil infrastructure agency involved in the case study in relation to the pre-cast sector were related to economic sustainability. The following effects of the lack of economic stability in the sector were identified (London and Chen, 2006):

- Cyclic loss of skill levels.
- High occupational and health safety occurrences.
- Low productivity rates.
- Poor product quality and high levels of remedial work.

- Wasted government resources to monitor a poor performing sector.

There is a dearth of literature and industry comment about the state of the pre-cast sector internationally and in Australia. Existing literature related to the pre-cast sector has largely concentrated on specialised and technical issues related to product performance such as product strength and quality (Yassin and Nethercot, 2007; Kwak and Seo, 2002). Consequently, economic performance and human issues including occupational health and safety (OH&S) and rework issues caused by a skill shortage within the industry have remained largely unexplored. Although the chronic shortage of skills in the construction industry (DEWR, 2005), and in particular the pre-cast sector, has been identified, it still remains to be researched.

The case study described in this chapter was aimed at exploring the problem of economic sustainability and associated unstable employment levels and high staff turnover of the Australian pre-cast sector using a supplier group strategy map. The supplier group strategy map is an approach that construction industry policymakers and procurers of capital works public sector clients can adopt to achieve sectoral change to develop productivity and performance innovation (London, 2007). There are five key activities: demand analysis; chain analysis; strategic alignment; supplier strategies; and organisational audit. A summary of these key elements is given later in this chapter.

The underlying assumption within the study is that groups of firms in sectors have a degree of influence over each other and it is this interdependency between the client and suppliers at various tiers that can improve or hinder the overall performance of the industry. However, governments have little real awareness of the intricacies of these interdependencies. This chapter reports the findings of the study, which attempts to reveal clients' perspectives of the pre-cast sector and of themselves and how key pre-cast agency suppliers perceive the civil infrastructure government agencies' actions and also their own sector (Figure 9.1).

The chapter is organised as follows:

- Discussion of construction supply chain economics theory.
- Description of the supply chain management blueprint and the supplier group strategy map.
- Description of the research methodology.
- Discussion of degree of implementation of state purchasing policy in relation to strategic procurement practices for pre-cast concrete case study by the case study organisation.
- Description of a process model for construction supplier group strategy map implementation.
- Implications for sectoral change and future research.

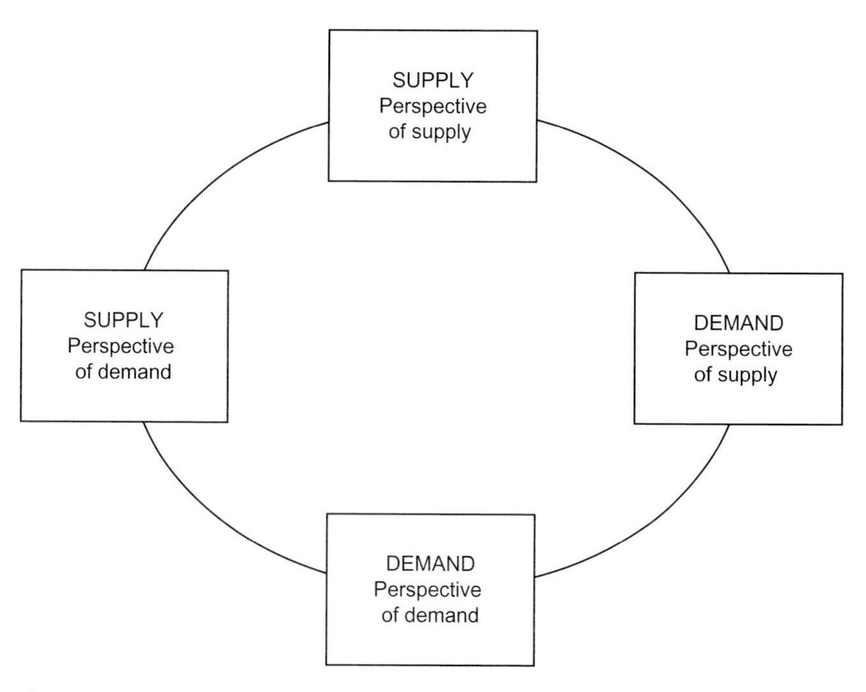

Figure 9.1 Supply (supplier)–demand (client) interactivity

Supply chain economics theory

> *The supply chain is constituted by firms involved in upstream and downstream contractual relationships, who deliver a commodity (product and/or service) related to the core business of firms delivering to projects. The supply chain, once formed, creates a flow of commodities, cash and information. The creation of the supply chain is impacted upon by the location of the individual firm within its market, which has unique economic structural and behavioural characteristics. The upstream and downstream linkages are affected by the nature of these markets. Linkages are also affected by the countervailing power relationships which occur between the customer and supplier markets at different tiers in the chain.* (London, 2004).

The supply chain management concept has gained the interest of the construction research community and policymakers through its successful implementation by manufacturing sectors, particularly to resolve firm performance problems. The general approach towards supply chain management to improve industry performance by policymakers has been through either of the following two model types (London, 2004):

1. *Normative managerial models*: where the aim is firm integration through collaborative relationships, based on an idealised homogenous industry, which is in fact fragmented and composed of numerous small to medium-sized firms.
2. *Positive economic models*: where the aim is improved economic customer and supplier performance in strategic relationships through the description, analysis and understanding of the market characteristics and drivers in relation to firm behaviour, based on the assumption of acceptance of an industry that is specialised and heterogeneous with varied structural and behavioural characteristics across individual markets.

It seems policymakers are implicitly or explicitly seeking positive economic models (London, 2004) for application yet many existing policies (development and/or implementation) and construction research are largely underpinned by implicit and/or explicit assumptions of industry homogeneity and have therefore tended to focus on normative managerial models. According to London:

> *. . . the greatest difficulty with supply chain management in terms of construction research theory and practical application is that currently too little is known about the structural and behavioural characteristics of chains and how to describe them.* (2004:1)

Therefore a positive economic model to aid mapping industry structural and behavioural characteristics was developed (London, 2004) which was used in this chapter. The model has been developed based upon past research in seven other major sectors. It highlighted structural and behavioural characteristics; however, explicit alignment with strategic industry objectives and performance measurement had yet to be explored. This study builds and develops this work to describe and map the underlying structures and behaviour specific to the pre-cast concrete market sector in order to explore an innovative supply chain management strategy for government.

Supplier group strategy map

Implementation at the sector level involves the development of a range of supply chain management strategies as indicated in Figure 9.2. As noted previously the blueprint was developed as a result of a previous study (London, 2004; 2007) and the supplier group strategy map was 'tested' and refined during research for this chapter.

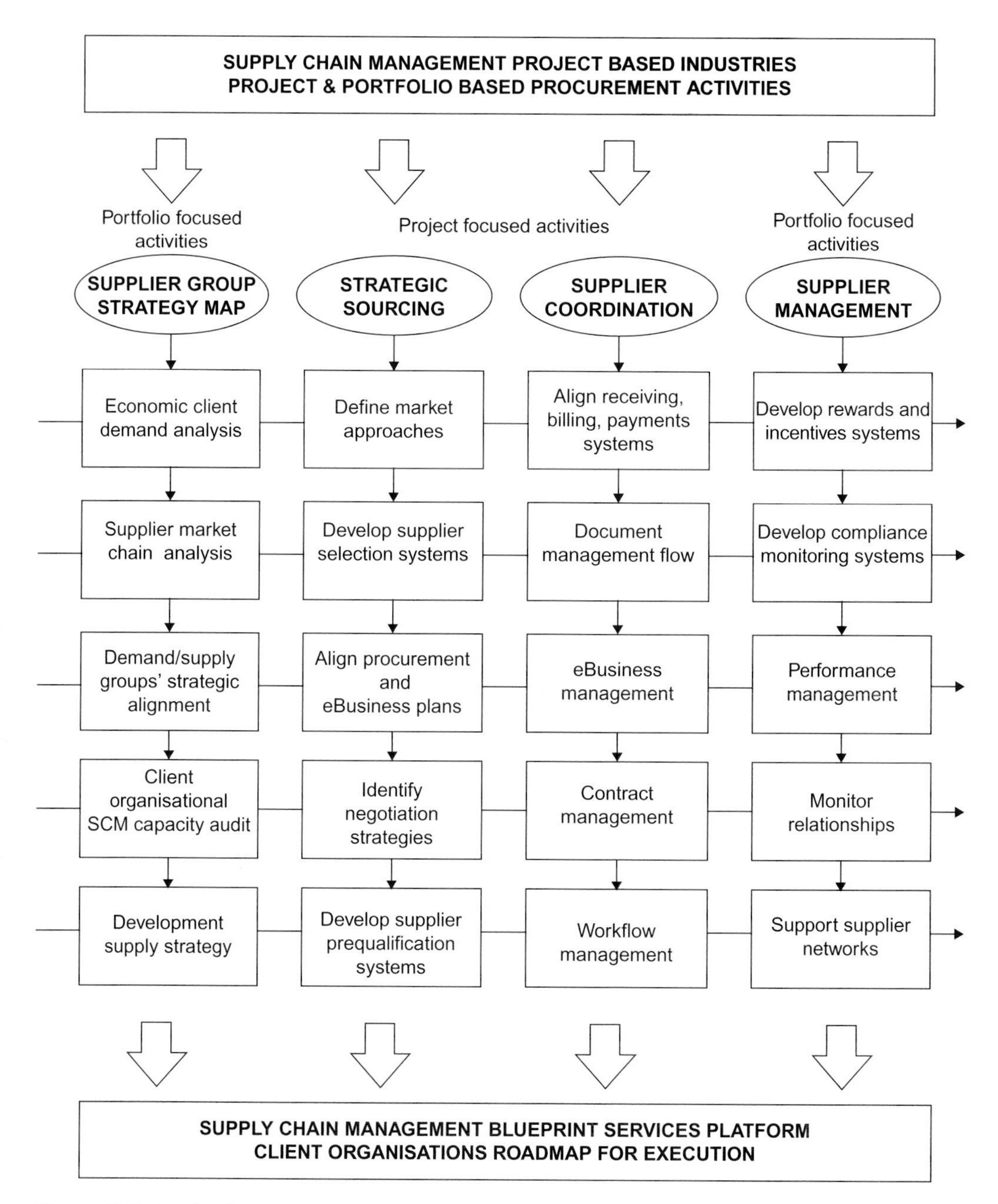

Figure 9.2 Supply chain management blueprint (London, 2007)

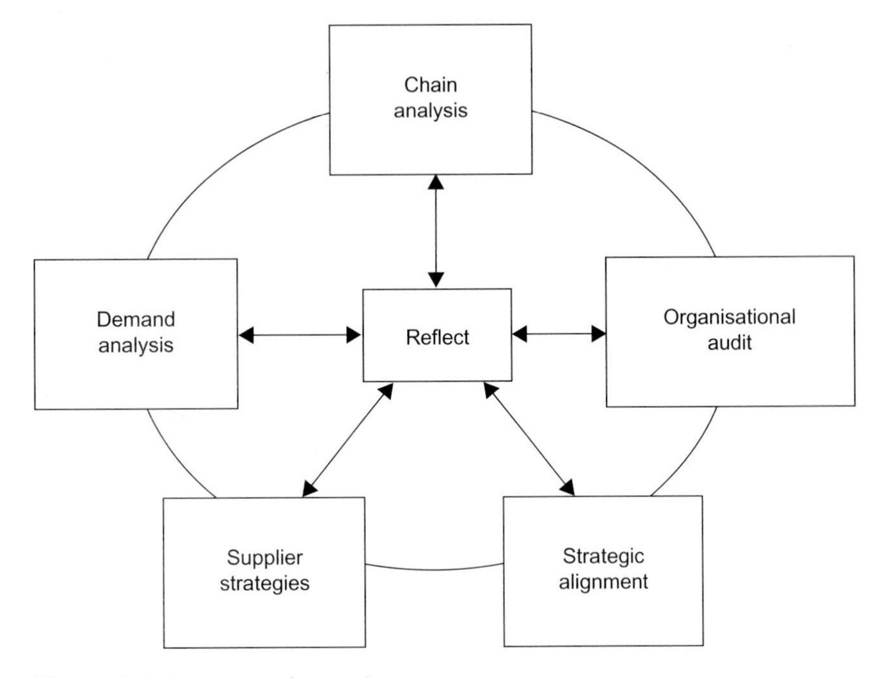

Figure 9.3 Five stages of a supplier group strategy map

The development of a supplier group strategy map for the two case studies, referred to as C&DW and PCC supply chains, will result from the following five activities (Figure 9.3):

- Demand analysis: purchasing history, future portfolio expenditure (regional/state), economic policy impacts upon the supply chains.
- Chain analysis: market characteristics, underlying structure and behaviour.
- Strategic alignment: development of key performance indicators for the supply chain.
- Supply strategy: risk versus expenditure categorisation of suppliers mapped against specific sourcing strategies.
- Organisational audit: internal business processes and policy.

One of the unique aspects to the map is the iterative nature (Figure 9.3) of the different activities within the project, whereby information gathered and learning and outcomes achieved at each stage should be continually applied to the next stage to improve practice. The different activities produce a series of events in a cyclic process. This study is aimed at exploring issues which emerge as the supplier group strategy map is put into practice by the agency, and understanding the extent that it can assist the agencies in supply chain and strategic procurement policy implementation.

Research methodology

An action research methodology is used for this study. Stringer (1996) describes action research as a cyclical process involving a 'look, think, act' routine (Table 9. 1).

Action research is qualitative and iterative, providing a flexible enquiry process carried out by individuals, professionals and/or educators within a professional practice to continually understand, evaluate and change in order to improve practice (Bassey, 1998; Dick, 2002; Frost, 2002; GTCW, 2002). Action research develops a rich understanding of complex issues and the subtleties of practice. It involves implementing actions which change existing programmes and practices, and the subsequent analysis of what happens (Rossman, 1998). When applied to the policy process, action research involves gathering and interpreting data to better understand an aspect of the policy process to learn and apply outcomes to improve practice (GTCW, 2002), namely exploring the process of supply chain and strategic procurement policy implementation and the difficulties associated with using the supply chain concept to improve policy, process and practices.

The study involved a collaborative approach between the research team and representatives from the state civil infrastructure agency of working through the supplier group strategy map and developing actions through understanding the nature of the problem from all stakeholders' perspectives. The research team facilitated the action based upon collecting the data and then presenting the findings to the agency representatives on the project. As the study progressed the findings were presented to a wider circle internal to the agency (more stakeholders). The actions were decided upon by the agency representatives in various combined workshops which involved the research team and

Table 9.1 Action research methodology (adapted from Stringer, 1996)

Stage	Key activities
1: Look	Gather relevant data/information/literature Build a picture and describe and define the situation
2: Think	Explore and analyse: what is happening here? Interpret and explain: how/why are things as they are? Refer to literature to make sense of findings
3: Act	Plan and determine actions Implement Evaluate
4: Document	Analyse and document outcomes

the representatives from the agency; rather than the usual situation where the actors design and conduct their own action(s).

A common difficulty associated with action research projects is the potential lack of response, disagreements and/or negative reactions from individuals or organisations that are being studied due to the high level of scrutiny placed upon work practices. This may lead individuals to be defensive and sometimes perceive that their knowledge and expertise is being challenged. Large organisations, including government agencies, can be a highly politicised environment and a difficult research site. Similarly researchers are intimately linked with the participants and the organisations and can find it difficult to evaluate the 'actions' as they are part of the development of the 'actions' as in this study. However, the benefit is access to rich data, information and insights.

Table 9.2 summarises the data collection techniques for the research described in this chapter based upon Stringer's (1996) model for action

Table 9.2 Research method

Stage	Key activities	Data collection
1: Look	Gather relevant data/ information/literature	Discussions on problem of the pre-cast concrete supply chain with agency representatives Discussion with state purchasing policy unit (2.5 hours' duration with three interviewees) on level of implementation of policy with civil infrastructure agency and attempt to identify examples of implementation in the property and construction industry Identification of unit of key implementation tool within the policy: risk versus expenditure and interactive website guidelines developed by unit for all other agencies
2: Think	Explore and analyse: what is happening here? Understanding/awareness of policy Discussion of early findings with reference to literature to make sense of findings	Focus group interviews: discussions with civil infrastructure agency on purchasing policy and risk versus expenditure tool to introduce ideas about supply chain management and then deeper discussions on the supplier group strategy map (four interviewees, 3 hours' duration) Individual semi-structured interviews (four interviewees, 1–3 hours' duration) Interviews with industry participants (eight interviewees, 1–4 hours' duration and site visits of yards) Internal agency interviews (11 interviewees, 1 hour in duration)
3: Act	Plan and determine actions Implement Evaluate	Not reported in this chapter
4: Document	Analyse and document outcomes	Not reported in this chapter

research. Table 9.2 only represents the specific formal events of data collection carried out on the project. Apart from these formal discussions, the project has involved 2.5 years of ongoing informal data collection discussions and events, whereby the research team has been progressively building up a picture and getting immersed in the case study agencies (for example, early discussions during the scoping of the project involving the case study agency representative transferring their knowledge of the supply chains to the chief investigator, project team meetings involving research team clarifying ideas with case study agencies, feedback discussions surrounding the analysis of findings, etc.).

Context of state strategic procurement

The understanding of strategic procurement began with a consideration of the state purchasing policy as requested by the agency responsible for construction industry policy development. Many of the ideas that were being put forward in the demand and chain analysis in the supplier group strategy map were principles which had already been documented in the state purchasing policy.

State purchasing policy unit

The first phase involved a discussion with various representatives and staff members from the Procurement Policy Unit who are responsible for whole of government purchasing policy. The unit is located within the agency responsible for construction industry policy development and is specifically responsible for the development of whole-of-government policy in relation to purchasing and assistance in implementation. A focus group interview with three employees was conducted. The research team's aim was to explain the research project and then to identify successes, experiences and insights in relation to the implementation of the policy directly with the pre-cast concrete sector or any construction sectors – particularly in relation to tools for implementation. The interview was recorded and transcribed and subjected to open coding, which involved the identification of themes and concepts as revealed by the individual interview transcripts (Miles and Huberman, 1994). The interview highlighted that the implementation of state purchasing policy by the civil infrastructure agency has been problematic in relation to the pre-cast sector and indeed in any construction sectors.

This phase also involved an explanation by the unit responsible for state purchasing policy development on the key elements of the policy and an introduction to one of the implementation tools; namely the risk versus expenditure tool and the associated interactive website. This tool is quite common in strategic purchasing and supply chain management literature.

The aim of these discussions was to locate exemplars of strategic procurement practice by government agencies associated with the property and construction industry so that the team could use these in exploring the supplier group strategy map with the civil infrastructure agency involved with the pre-cast concrete sector. No exemplars were provided for benchmarking against the pre-cast concrete supply chain which was to be the focus of this research. It was highlighted in those discussions that there had been little implementation of strategic procurement principles by the construction agencies that they could identify and provide to the research team. The research team then examined the risk versus expenditure tool and the website as just one tangible example of what the policy had to offer and took this to the next series of interviews and discussions with the civil infrastructure agency. A consideration of the summary document of the state purchasing policy and the website provided by the unit informed the future interviews.

The following key themes were identified at this stage:

- Examples of implementation in other sectors based upon leverage purchasing (i.e. high volume) could be identified.
- Examples of implementation in other sectors based upon purchasing of 'simple' products and/or services (e.g. stationery, IT) could be identified.
- Construction sector implementation examples could not be identified.
- There was lack of awareness and understanding of construction sector complexities by the unit.
- There was frustration at the lack of uptake of various tools, specifically, for example, the Boston Map (Supply Positioning), a key tool used to assess risk versus expenditure for strategic procurement (Figure 9.4).
- The Boston Map (Supply Positioning) tool is suggested for use by all government agencies and departments in the state purchasing policy:

> *'. . . using a procurement management tool called supply positioning, goods and services are plotted according to their relative expenditure and difficulty in securing supply. This is a good way to determine where the procurement effort should be focussed in the Corporate Procurement Plan for the year.'* (QDPW, 2001).

The research team could, however, provide an exemplar of how to use the Boston Map from past research (London, 2004) from a large construction material supplier's perspective. Although not an ideal

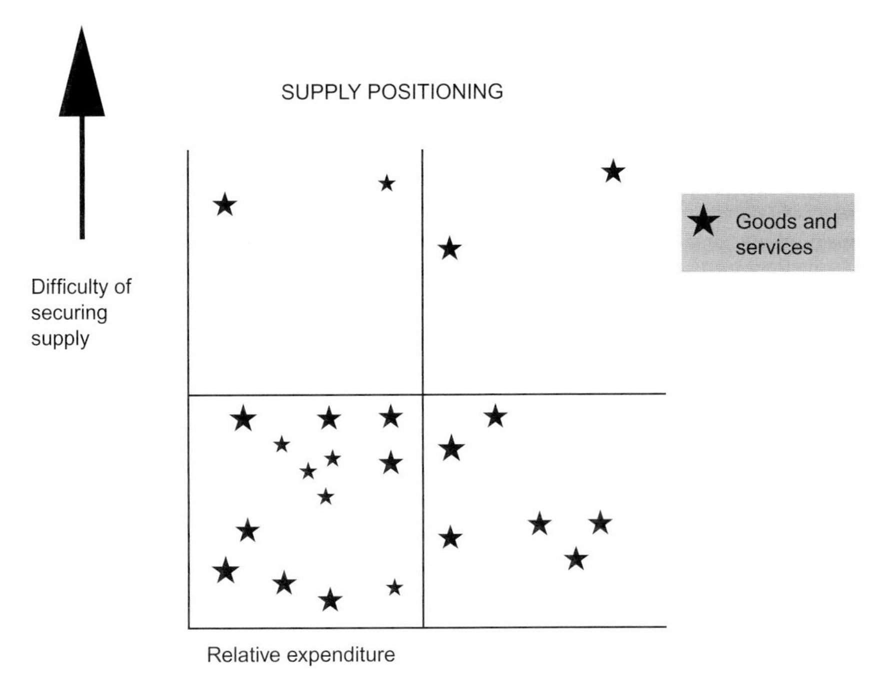

Figure 9.4 Boston Map (supply positioning tool) (QDPW, 2001)

example because it is the use of the tool from a non-public sector organisation and also from a material supplier perspective rather than from a construction project client perspective, it highlighted the relationship of risk versus expenditure to the type of relationships that could be developed.

This phase of the research served as a first step towards identifying the context of state purchasing policy. The next stage of the analysis involved identifying the degree of implementation of the policy within the agency involved with the pre-cast concrete sector. It is noted that the state purchasing policy is not a policy on supply chain management. Supply chain management has many more activities and is a much more complex environment than the purchase of an individual commodity. There are many more policies which the civil infrastructure agency is required to implement and some specifically designed for procurement within the construction industry which are quite context specific.

At this stage a consideration of the theory in relation to policy development and implementation was undertaken. It is noted that despite achievement of various steps in the policy process cycle, policy implementation and associated outcomes cannot always be guaranteed as the practice of policy development and implementation is a complex and multi-faceted matrix of politics, policy and administration (Bridgman and Davis, 2004).

The Boston Map (Supply Positioning) tool provides four segments and they are often referred to as tactical purchasing (low expenditure/low risk/difficult); leverage (high expenditure/low risk/difficult); strategic security (low expenditure/high risk/difficult); and strategic critical (high expenditure/high risk/difficult).

The next stage of the research involved a focus group interview with the civil infrastructure agency employees.

Difficulties with implementation of risk versus expenditure tool

A focus group interview was conducted with internal staff members involved in project and contract management and policy development in relation to the pre-cast concrete sector from the civil infrastructure agency. One of the aims of the interview was to identify levels of awareness and understanding of the state purchasing policy and the various associated support documentation including the guideline, manual and website and specifically the risk versus expenditure tool (Boston Map). The staff members present at the workshop included a senior policy manager, two senior engineers and one inspector.

Although considered reasonably simple and effective as a first step towards determining approaches towards suppliers, this tool had never been used before by the case study organisation interview participants, despite it being an integral and significant part of state purchasing policy. The manner in which the participants reacted to the procurement management tool suggested it was too generic, and therefore unsuited to the different characteristics of players within the construction industry and the pre-cast concrete sector. Lack of awareness, understanding and perhaps guidance on how to use the tool appropriately for the civil construction sector, and perhaps specifically the pre-cast sector, resulted in the participants dismissing the entire policy, policy manual and associated guidelines and the agency that had developed it. Our discussions in the focus group interview then progressed to what was wrong with the tool and how it could be enhanced. Contributions were particularly useful from the senior engineer and they are highlighted in the following text. The positioning of the agency within the supply chain, and indeed the changing position, added a layer of complexity to the risk dimension, as demonstrated by the reactions of the participants:

> *There's long-term risks and short-term risks. With a pipe, there's the short-term risks where if the pipe doesn't fit then we could just get another one and refund it . . . then there's the long-term risk where if in 20 years time the pipes are no good then we've got to dig it up and put in a new one.* (Senior engineer 1)

> *Because of where we sit in the industry and on the supply chain . . . We're not just buying or selling, we're buying, building and owning . . . so there are different levels of risks . . .* (Senior engineer 1)

More specifically, regarding the use of the tool it was highlighted that it could perhaps be modified through a closer examination of the risk dimension:

> *I'm just wondering if there's another dimension to the risk part of it . . . as an owner, there just may be another dimension to it . . . may be it needs to be tailored to this particular project . . . (Senior engineer 1)*

The civil infrastructure staff members described difficulty of securing supply as their exposure to various elements of risk of supply of pre-cast concrete. They then suggested that risk and expenditure were not static but changed over time. They indicated that the Boston Map is useful only if it reflects the more complex real world situation. The researchers then developed a series of Boston Maps to more accurately reflect the information provided during the interview, incorporating the changing nature of the relationship between risk and expenditure. Figure 9.5

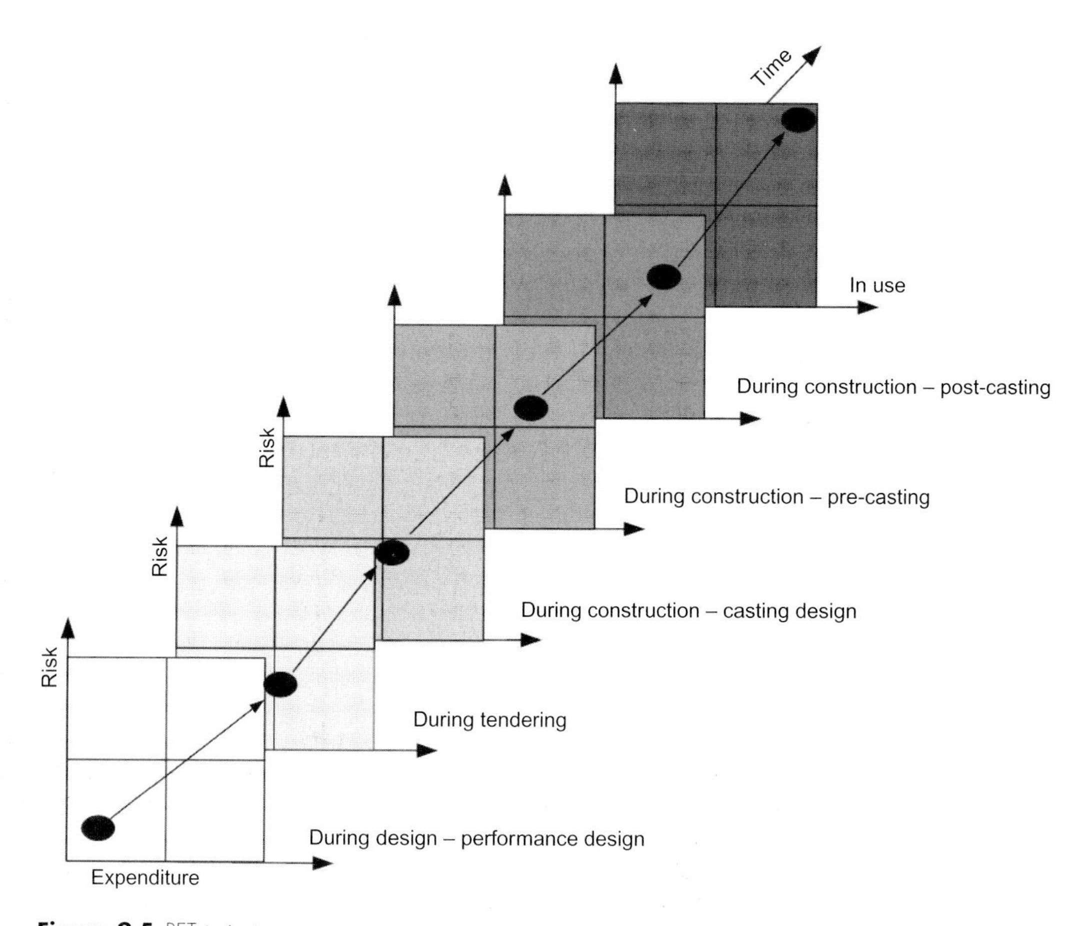

Figure 9.5 RET trajectory

relates the three key dimensions of risk, expenditure and time. For example during the design stage an error identified internally on a drawing by a designer or consultant was considered to be of lower risk in comparison to an error identified during construction in casting design. Then if an error is identified during construction post-casting it is considered to be an even higher risk. The highest level of risk to the agency occurs when an error is identified in use. The level of expenditure required to rectify the error, and thus the risk of failure in relation to the supply of the item, correspondingly increases with time. We can locate suppliers in one of the four segments but over time this may shift. The most significant point is that risk and expenditure may change over time, and therefore there can be a risk and expenditure trajectory.

The original intention of the Boston Map specifically for strategic procurement, however, is to map the financial risk to the purchaser of securing supply relative to the volume of purchasing; thus indicating the nature of power that the purchaser has in relation to the supply market and how vulnerable they are in relation to supply. It was difficult for the participants to consider financial risk without considering their overall risk and other dimensions of risk in relation to the supply of pre-cast concrete. For example technical risk of failure and identification of errors at different times of the asset lifecycle exposes the client to different levels of financial risk and different levels of risk of securing supply depending upon the nature of the failure, complexity level of product, current market conditions and the nature of the project contract (procurement strategy and contract type).

A further illustration of how problematic policy implementation is for these participants is provided by the nature of the pre-cast concrete sector. The sector is complex in two main ways; it is complex in relation to the actual product and also in relation to the processes and people involved in the sector. Pre-cast concrete products are either standardised or unique. Services are provided especially concerning the unique products where design is included. Therefore supply can be complex, involving a high level of design and construction knowledge and capability. Typically the civil infrastructure agency policy is for high standards in relation to design, construction and materials, leading to low maintenance of civil infrastructure during its whole lifecycle. High standards also lead to safer structures which is another key policy objective of the particular civil infrastructure agency involved in this case study. Secondary objectives include maintaining a stable workforce, a safe working environment in the pre-casting yards and a sustainable flow of projects to the sector. However, the state purchasing policy is primarily concerned with effective and efficient use of government resources in relation to government spend, but what this constitutes in the construction industry and pre-cast sector can only be determined when more specific objectives are established to inform implementation. Generic purchasing policies are an important part of supply chain management but additional strategies may be needed as they may be insufficient for complex projects and their supply chains.

A conflict may lie in the approaches taken by the two agencies involved in the case study in that the purchasing unit's whole of government objectives fundamentally seek to amalgamate the individual purchases and that is their starting point; whereas the philosophical starting point of the civil infrastructure agency is typically that the project is an individual, yet major, 'purchase'. The civil infrastructure agency does not appear to consider government purchasing objectives holistically across their portfolio of projects and the efficient and effective use of their internal resources. For example, although there is a strategic procurement unit and a structures unit within the civil infrastructure agency, who both interact with the pre-cast sector, they have still not aligned risk, expenditure and the various internal business processes necessary to efficiently and effectively manage their various interactions with the sector.

One policy objective is to achieve a good quality product that 'lasts 100 years'. The consequence of the shortfall in performance improvements in the sector became clear through the research. For example:

> *Most of the industry's focus is always on the short term, can we build it fast, make more money and get it done quicker. Whereas when it comes to concrete structure by cutting off the hydration process you're making something that lasts 100 years last 50 years because you didn't do it right to start with.* (Senior engineer 2)

Implementing agencies require solutions that are specific to the problems of the respective sectors and perhaps even specific projects. The understanding of sector and project specificity is largely absent from the broad objectives of the agency that developed the policy. This is not intended to be a criticism, as the policy is a generic policy and each agency is required to tailor it to their individual needs. In other words, for the effective utilisation and implementation of policies there is a need to 'tailor' the policy to the construction industry. When the agency that had developed the policy was interviewed, a clear desire emerged to assist and develop deeper penetration of the policy into the particular implementing agency involved in this case study. However, as noted previously there were no successful implementations of this purchasing policy across the entire state in the property and construction industry, even though the industry represented the bulk of government sector purchasing activity. It should also be noted that not only has there not been any successful implementation of this purchasing policy in the civil infrastructure case study agency, the agency staff members also lacked awareness in relation to the concepts of strategic purchasing and supply chain management. It is noted that the agency staff members interviewed in the other case study, which is not described in this chapter, also were similarly unaware of the policy and its meaning.

This discussion has demonstrated the types of issues confronted in practice for only one specific sector of the construction industry, albeit a critical supplier group for the civil infrastructure agency. This study has limitations in that it is focused on only two government agencies through the two case studies. The subsequent interviews (see Table 9.2) with the civil infrastructure agency which are not reported on in this chapter supported the contention that there was a general lack of diffusion of the principles of strategic procurement espoused in the policy. The policy implementation may well be widespread across all other implementing agencies. These findings are now discussed in context of the literature.

Policy implementation discussion

Most theoretical models of the policy process are largely based on the assumption that the organisations or government agencies responsible for implementing policies have simple machine-like characteristics; that is, required actions are deemed to be rational based upon simplistic chains of cause and effect (Ryan, 1996; Wilkinson, 1997). There is a failure within these models to recognise that, in actual practice, government agencies do not necessarily implement policies based on these strictly rationalised models (Goggin *et al.*, 1987, 1990; Mazmanian and Sabatier, 1989).

The key barriers to the successful implementation of policies in the real world environment that have been identified in the literature include:

- Conflicting objectives and directives at different levels of government, agencies and/or implementing actors.
- Limited competence.
- Incomplete specification.
- Insufficient resources.

Each of these will now be explained in further detail. First, policies developed by one level of government do not necessarily take into account the objectives of another level. The same applies for agencies, whereby implementation can vary considerably across different implementing agencies. Agencies seek specific solutions to particular problems. Secondary evidence has shown that some agencies do better than others in implementing the same programme (Pressman and Wildavsky, 1973; Gunn, 1978; Beer *et al.*, 1990; Hasenfeld and Brock, 1991; Goggin *et al.*, 1987, 1990).

Second, some policies fail because implementing agencies lack the necessary capacity, expertise or commitment to implement them (Fenna, 2004). For example, when political objectives do not match their capacity, this provides opportunity for implementing officers in the agencies to ignore official instructions and pursue personal policy preferences. Thus we might see policy documents and guidelines which literally 'sit on shelves' and are never used or explored further. We saw evidence of this in our study. Another explanation is that agencies may not actively resist implementation but instead, because they hold different assumptions and perceptions of what the policy means, may in fact believe they are implementing the policy.

Third, some government policy documents set out broad general principles with insufficient and/or unclear information to guide policy implementation (Patton and Sawick, 1993; Blackmore, 2001; Fenna, 2004). A key problem with diversity and density of guidance is policy ambiguity, which offers either the potential for implementing agencies to exercise informal discretion (Hill, 1969) or the risk that implementing officers can be attempting to achieve conflicting objectives simultaneously hence leading to the lack of implementation (Blackmore, 2001).

Fourth, policy implementation requires adequate resource availability and control (Mazmanian and Sabatier, 1989; Winter, 1990; Anderson, 1994; Fenna, 2004). It appears that adequate resources are available for policy implementation with the formation of the strategic procurement unit within the civil infrastructure agency; however the unit was only recently formed (albeit some 9 months prior to the interview). Failure to emphasise the significance of specific implementation tasks can also result in inadequate resources allocated to enforce or support programme objectives, which in turn prevents the careful and thorough implementation of policy.

Despite the best of intentions by the state purchasing unit through extensive development and provision of policy documents to guide policy implementation there was still a lack of diffusion of the policy and a lack of widespread understanding of key strategic procurement principles in the two case study agencies at strategic and operational level and this is perhaps an important point to consider. Although the strategic procurement units within the agencies understood the key strategic procurement and supply chain management principles, other staff members who did not come from a purchasing background but were from a 'hands on' construction background, lacked an understanding of the principles, particularly the economic principles which underpin the demand and chain analysis in the supplier group strategy map.

Transitional space

The concept of implementation gap is not particularly new, for example as seen in Australia in the early 1990s when a comprehensive and fairly well harmonised policy 'Total Asset Management Framework' was developed and adopted by the states and the federal government. It was some 10 years before any real penetration of that policy came into effect. During that time the policy document 'sat on the shelf' as construction professionals within agencies often struggled because of lack of time and lack of guidance on contextualisation of the policy specific to their needs. The theory also supports the identification of an implementation gap (Barrett and Fudge, 1981) and the present research builds upon this by exploring further the concept of 'transition space' as a potential solution towards addressing the gap.

There is a transitional space between policy developed by one agency and then implemented by another agency, which needs to be managed. The context of both the agency responsible for development of generic policy and the agency responsible for implementation in a specialised environment should be considered to determine who should take responsibility for implementation. Specific policies should be developed in accordance with operational context, considering internal factors, such as core professional workforce capability, capacity, organisational culture and outlook, and external factors, such as industry characteristics. It is not simply enough to identify that an implementation gap exists and identify the barriers inhibiting policy implementation. We need to begin to provide a little more substance to the characteristics of this 'gap', and what could be involved in bridging this gap and moving towards greater levels of implementation of the policy.

Therefore four key impediments have been highlighted which are generic to numerous other studies. The reason why implementation of strategic portfolio concepts related to procurement are not adopted by the particular civil infrastructure agency involved in this case study was not a specific focus of the study. The purchasing policy is not a supply chain management policy; strategic purchasing is one aspect of supply chain management. The case study served to highlight the difficulties of implementing a policy which is not central to the operations levels of an agency, i.e. project and contract management of state civil infrastructure. The civil infrastructure agency also has roles and responsibilities in relation to industry development and therefore has objectives which need to consider the long-term state of construction markets to support business development. At times this may not conflict or not completely align with the purchasing policy objective of efficient management of government resources.

The concept of 'transition space' is identified as an approach to bridge an implementation gap for an innovative supply chain management policy. Taking these issues into account a process model for policy development and implementation is proposed for the development of a supplier group strategy map for the property and construction sectors to address the property and construction context.

Process model

A policy cycle model (Figure 9.6) is used by the Australian system of government to ensure that the process of policy development, implementation and evaluation matches the broader objectives of governments (Bridgman and Davis, 2004). The policy cycle model offers a series of generic activities to guide better practices towards successful policy implementation. Figure 9.7 illustrates a model which aims to align theory to practice in relation to public sector construction supply chain policy and practices through an integrated supply chain policy approach. There is no explicit construction supply chain management policy available and this is a conceptual model based upon the difficulties identified in this case study. The integrated construction supply chain implementation process model (Figure 9.7) builds upon the generic policy cycle model (Figure 9.6) to provide increased detail and relevance specifically for construction supply chain policy implementation by the civil infrastructure agency.

A supply chain management policy would require integration of a number of existing policies involving government state purchasing

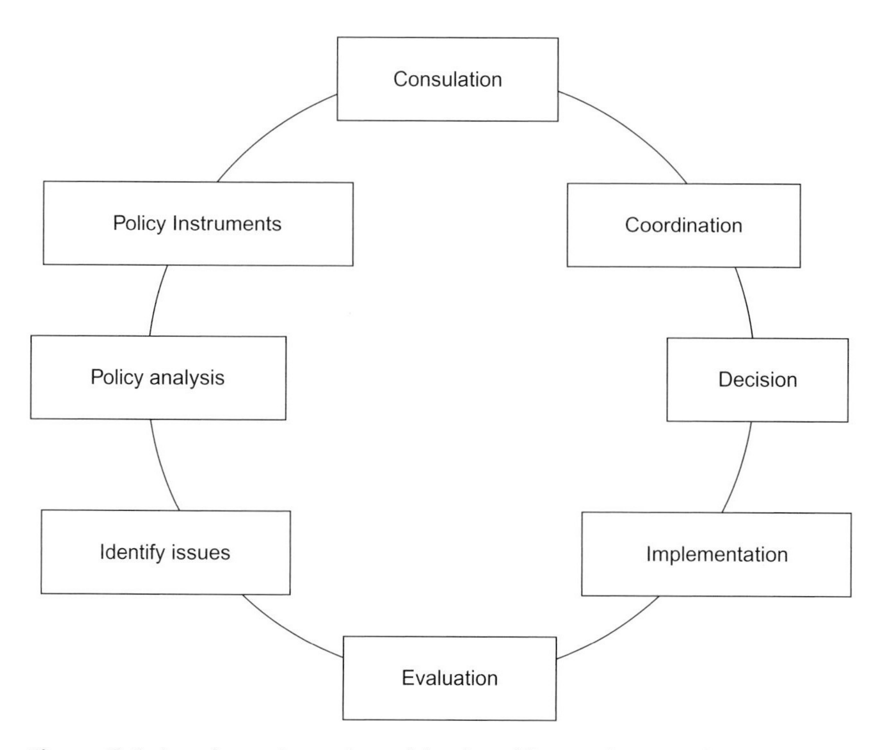

Figure 9.6 Australian policy cycle model (adapted from Bridgman and Davis, 2004)

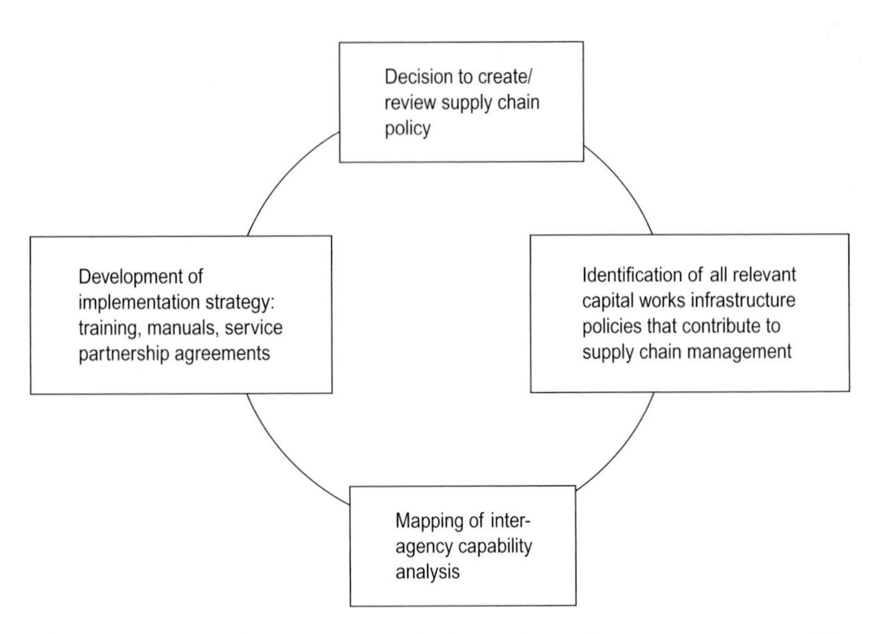

Figure 9.7 Integrated construction supply chain policy implementation process model

policy and a number of other domains which affect supply chain management; these have been highlighted in the supply chain management blueprint strategy (see Figure 9.2). The process model is based upon the premise that a deeper understanding and explicit recognition of the different roles and responsibilities that various agencies play can contribute towards increased diffusion of supply chain management in the implementing agency at strategic and operational level. The first step involves an identification of all relevant capital works civil infrastructure policies that contribute to issues related to supply chain management which could include tendering, strategic procurement, equity, probity, information management, etc. As the implementation of supply chain policy involves a range of policies, there is a number of agencies who will have expertise and who can provide input at various points. Therefore the expertise of the various agencies needs to be mapped to identify the core civil construction expertise as well as the specific expertise that will contribute to supply chain management. Following the mapping of inter-agency capability analysis a strategy for implementation needs to include service partnership agreements across various identified agencies which could enhance inter-agency capability. Various guidelines and training programmes could be considered with tools for implementation that are relevant.

The research has shown that there is a lack of explicit information available to all the parties. There were indications of an implementation gap in the early discussions with the state purchasing policy unit which was confirmed as the action research proceeded. The standard risk versus expenditure model, the Boston Map, was not adequately understood nor could be applied by the civil infrastructure implementing agency. This is because the policy did not capture the complexities and specialised nature of the agency, nor capture the industry context. It is noted that other implementing agencies have adopted the principles and therefore the comments are specific to the particular case study described in this chapter. It is noted that it was not the intention of the purchasing unit to develop tools for context-specific application.

An important component of the process model is the explicit identification of the governmental capacity to develop and implement policy across agencies. Table 9.3 takes the process model down to a greater level of detail mapping at the level of agency capability to support policy development and implementation. The inter-agency capability analysis was developed based upon an interpretation of the participants' description and discussion of their role and activities. This inter-agency capability analysis can assist in the development of service level agreements between the various agencies. It is important that across agencies and within agencies the contributions are made explicit, to enable appropriate levels of resources.

Conclusions and recommendations

One advantage of action research is that the organisations involved can benefit from the research findings. The case study organisations identified the need for strategic procurement units to manage the transition space. The aim of these units is to bridge the gap by addressing the complexity of the agencies and sectors within civil construction and procurement policy making. It is recommended that all parties address:

- Context-specific policy.
- Directly applicable tools related to the specific organisations.
- Identification of roles to develop policies, strategies, tools and techniques, which are meaningful and context specific.

Development of homogeneous or generic policies provides a starting point, yet it is problematic as the implementation level by government agencies and for industry context is limited. The problem lies in what has already been identified as the policy 'implementation gap'. To overcome the negative connotations and limited direction that the implementation gap brings to the discussion we suggested building upon past theory. The gap should be viewed as a 'transition space' which

Table 9.3 Inter-agency capability analysis

Government agency	Role	Skills, knowledge, experience, expertise				
		Sector-specific operational knowledge	Construction industry operational knowledge	Project procurement	Construction policy	Strategic procurement policy
Procurement and Building Agency						
Procurement Policy Unit	Develop whole-of-government purchasing policy	Low	Low	Low	Low	High
Civil Infrastructure Agency						
Strategic Procurement Unit	Implement State Purchasing Policy	Low	Low	High	Low	High
Capabilities & Delivery: Policy	Provide auditing services to Structures Unit	Medium	High	Low	High	Low
Structures: Concrete Technology	Purchase pre-cast concrete products	High	High	Medium	Low	Low

needs to be managed. This is a specific problem for the agencies involved in this action research. However, it raises wider lessons for any organisations and in all supply chains. The policy, market and organisational contexts have to be taken into account. General policies and strategies must be made specific to work. Failure to do so may result in a failure of supply chain strategies and procurement policies. It will certainly inhibit collaborative working and may create the opposite effect. The authors suggest that improved collaborative relationships within and across agencies are required, as well as explicit identification of roles, responsibilities, skills, knowledge and capabilities to support service level agreements between agencies responsible for policy development, policy implementation assistance and policy implementation.

Australian government purchasing policies, which are highly supportive of supply chain economic principles, need to be merged with the various current construction project civil infrastructure procurement policies, which are typically project focused and do not adequately consider a supply chain approach. The supply chain blueprint (London, 2007) provides a raft of strategies which attempts to merge leading thinking in civil construction with supply chain management and economics complementing the findings reported in this chapter.

Acknowledgement

The research described in this chapter has been performed by the Australian Cooperative Research Centre for Construction Innovation.

References

Anderson, J. (1994) *Public Policymaking: An Introduction*, 2nd Edition. Houghton Mifflin, Boston.

Barrett, S. and Fudge, C. (1981) *Policy and Action: Essays on the Implementation of Public Policy*. Methuen, New York.

Bassey, M. (1998) Action research for improving educational practice. In: Halsall, R. (ed.) *Teacher Research and School Improvement: Opening Doors from the Inside*. Open University Press, Buckingham.

Beer, M., Eisentat, R. and Spector, B. (1990) *The Critical Path to Corporate Renewal*. Harvard Business School Press, Cambridge.

Blackmore, M. (2001) Mind the gap: exploring the implementation deficit in the administration of the stricter benefits regime. *Social Policy & Administration*, 35(2), 145–162.

Bridgman, P. and Davis, G. (2004) *The Australian Policy Handbook*. Allen and Unwin, NSW.

Cox, A. & Townsend, M. (1998) Strategic Procurement in Construction: *Towards Better Practice in the Management of Construction Supply Chains, Vol. 1*, 1st Edition. Thomas Telford Publishing, London, UK.

Department of Employment and Workplace Relations (DEWR) (2005), National Skills Shortage List – 2004. www.workplace.gov.au, accessed Oct 2005.

Dick, B. (2002) Action research: action and research. www.scu.edu.au/schools/gcm/ar/arp/aandr.html, accessed Aug 2007.

Fenna, A. (2004) *Australian Public Policy*. Frenchs Forest, Pearson Education, NSW.

Frost, P. (2002) Principles of the action research cycle. In: Ritchie, R., Pollard, A., Frost, P. and Eaude, T. (eds.) *Action Research: A Guide for Teachers. Burning Issues in Primary Education*, Issue No. 3. National Primary Trust, Birmingham.

General Teaching Council for Wales (GTCW) (2002) *Professional Development Pilot Projects*: Information Booklet 2002–2003. GTCW, Cardiff.

Goggin, M., Bowman, A., Lester, J. and O'Toole, L. (1987) *Policy Design and the Politics of Implementation: The Case of Child Health Policy in the American States*. University of Tennessee Press, Knoxville.

Goggin, M., Bowman, A., Lester, J. and O'Toole, L. (1990) *Implementation Theory and Practice: Towards a Third Generation*. Scott Foresman and Company, Glenview.

Gunn, L. (1978) Why is implementation so difficult? *Management Services in Government*, 33, November, 169–176.

Hasenfeld, Y. and Brock, T. (1991) Implementation of social policy revisited. *Administration and Society*, 44(4), 451–479.

Hill, M. (1969) The exercise of discretion in the National Assistance Board, *Public Administration*, 47, Spring, 75–90.

Kenley, R., London, K. and Watson, J. (2000) Strategic procurement in the construction industry mechanisms for public sector clients to improve performance in the Australian public sector. *Journal of Construction Procurement*, 6(1), 4–19.

Kwak, H. and Seo, Y. (2002) Shrinkage cracking at interior supports of continuous pre cast prestressed concrete girder bridges. *Construction and Building Materials*, 16(1), 35–47.

London, K. (2004) *Construction Supply Chain Modelling*. PhD thesis, Faculty of Architecture, Building and Planning, University of Melbourne.

London, K. (2007) *Construction Supply Chain Economics*. Taylor and Frances, Oxford.

London, K. and Chen, J. (2006) Construction supply chain economic policy implementation for sectoral change: moving beyond the rhetoric. *Proceedings of RICS Cobra Conference*, 7–8 September, UCL, London.

Mazmanian, D. and Sabatier, P. (1989) *Implementation and Public Policy*. University Press of America, Lanham.

Miles, M. & Huberman, A. (1994) *Qualitative Data Analysis*, 2nd Edition. Sage Publications, Thousand Oaks, USA.

Patton, C. and Sawick, D. (1993) *Basic Methods for Policy Analysis and Planning*, 2nd Edition. Englewood Cliffs, Prentice Hall, NJ.

Pressman, J. and Wildavsky, A. (1973) *Implementation: How Great Expectations in Washington are Dashed in Oakland*, 3rd Edition. University of California Press, Berkeley.

Pryke, S.D. and Pearson, S. (2006) Project Governance: case studies on financial incentives. *Building Research and Information*, 34(6), 534–545.

Queensland Government Department of Public Works (QDPW) (2001) *Corporate Procurement Planning: Better Purchasing Guide*. www.qgm.qld.gov.au, accessed Aug 2007.

Rossman, G. and Rallis, S. (1998) *Learning in the Field: An Introduction to Qualitative Research*, 2nd Edition. Sage Publications, Thousand Oaks.

Ryan, N. (1996) A comparison of three approaches to programme implementation. *International Journal of Public Sector Management*, 9(4), 34–41.

Stringer, E.T. (1996) *Action Research: A Handbook for Practitioners*. Sage Publications, Thousand Oaks.

Walker, D., Hampson, K. and Peters, R. (2002) Project alliancing vs Project partnering: a case study of the Australian National Museum project. *Supply Chain Management*, 7(2), 83–92.

Warren, M. (1993) *Economics for the Built Environment*. Butterworth Heinemann, Oxford, UK.

Wilkinson, D. (1997) Whole system development – rethinking public service management. *International Journal of Public Sector Management*, 10(7), 505–533.

Winter, S. (1990) Integrating implementation research. In: Palumbo, D. and Calista, D. (eds.) *Implementation and the Policy Process: Opening up the Black Box*. Greenwood Press, New York.

Yassin, A. and Nethercot, D. (2007) Cross-sectional properties of complex composite beams. *Engineering Structures*, 29, 195–212.

10 Construction and women

A comparison with the medical profession

Dilanthi Amaratunga, Menaha Shanmugam, Richard Haigh and David Baldry

Background

The construction industry is traditionally considered a white, male-dominated industry. There are over 11 million women employed in the UK, accounting for almost 50% of the workforce (Fielden *et al.*, 2000). According to the Construction Industry Training Board (2003), women only account for 9% of the construction workforce. This means construction continues to be a male-dominated industry. The concentration of men and women in different kinds of jobs, which could be described as gender or occupational segregation, conforms to societal expectations of the gender. Accordingly, women are more likely to hold administrative and secretarial positions whereas men are more likely to work in manufacturing and production. This gender segregation damages the UK's economy and competitiveness by contributing to skill shortages and the gender pay gap, and also restricts individual choices. Further, the predominant image of construction is described as a male-dominated industry requiring brute strength and a good tolerance for outdoor conditions, inclement weather and bad language (Agapiou, 2002). It is principally this image that makes the women uninterested in the industry. There is a little realisation that the industry is becoming high-tech and that mental strength and commitment are more needed than so-called brute strength. In this context the low number of women in the industry shows the under-utilisation of human resources based on gender patterns. It could therefore be said that women are under-represented.

One of the major reasons for the under-representation of women in construction is that women are confronted by a significant number of barriers. From a literature survey the major barriers have been identified as:

- The image of the industry (Fielden *et al.*, 2000, 2001; Bennett *et al.*, 1999; Gale, 1994).

- Lack of career knowledge (Harris Research Centre, 1989; Fielden *et al.*, 2000; Agaipou, 2002; Gale, 1994).
- Culture and working environment (Bennett *et al.*, 1999; Dainty *et al.*, 2000; Fielden *et al.*, 2000).
- Male-dominated training courses and recruitment practices (Fielden *et al.*, 2000, 2001).
- Family commitments (Greckol, 1987; Fielden *et al.*, 2000; Agapiou, 2002; Lingard and Francis, 2002; Lingard and Lin, 2004).

The report *Accelerating Change* led by Sir John Egan (1998) expressed concern regarding the shortage of people with the technical and management skills to fully utilise the new techniques and technologies available. UK construction demand is high for the decade and the industry is suffering from skill shortage in both craft and manual trades, and at the professional level (Whittock, 2002). In UK, the construction industry has the second highest level of skill shortages as a percentage of total vacancies. This skill and labour demand may be a threat to the long-term growth of the industry and it may also challenge the industry's capability to deliver the projects on time, within the budget and at the desired quality. The UK construction industry is facing recruitment problems with its traditional source of labour: young men aged 16–19 (Gurjao, 2006).

The constant reliance on a limited recruitment base disadvantages the industry by disregarding half the population and the diversity of skills these people have to offer. By restricting the possible workforce, the industry is limiting the choice of applicants at its disposal, which in turn may lead to the recruitment of lower-quality employees. Indeed, it is said that a major obstacle to the industry to recruit the best people is the fact that half of the population is largely ignored by the industry (Green, 2005). A study by Green (2005) further highlights the current position: *'it's a pretty rare breed of woman that works in the industry'* (p.58); *'we need more women to fill the skills gaps and to make a change'* (p.46). Currently, construction employers recruit and rely increasingly on workers from overseas, either inside or outside the European Economic Area (EEA), giving rise to immigration issues with an increasingly diverse force (Gurjao, 2006). However, increasing the number of women in the construction workforce would also contribute to solving the skill shortage problems to a certain extent. Thus the recruitment of women is imperative to achieving these objectives and prolonging the growth of the industry.

The construction industry is not homogeneous; it comprises a wide range of multi-disciplinary activities. It also involves various stakeholders which naturally create a fragmented culture within the industry. Therefore, promoting collaboration and team work are crucial in order to reduce the fragmentation and increase integration among stakeholders. A review of recent relevant social science-based research revealed that women play an important role in promoting collaboration and team work as they typically adopt nurturing qualities. Female values generally tend to be developed through socialisation processes that include building relationships, communication, consensus building, power as influence, and working together for a common purpose (Trinidad and Normore, 2005). Dainty *et al.* (1999) identify that increasing the number of women working in construction may go some way to improve this current status of the industry, firstly by utilising the full range of skills available in the population and secondly by assisting construction organisations to become more efficient and adaptable to the needs of its customers. Similarly, research by Bennett *et al.* (1999) identified that women are of benefit to construction organisations as by their nature they are good with people, less confrontational and are more likely to listen to the opinions of others, which will be beneficial when dealing with clients. Studies by Gale (1994) also acknowledged that these 'feminine traits' were beneficial in negotiation situations. Research by Agapiou (2002) found that more women working in construction will better the industry's image and go some way to improving the current skills shortage by aiding the recruitment of others, not just women but also other under-represented groups, and men who previously may not have considered a career in construction. Despite these advantages, the number of women working in the industry is constantly low. The barriers faced by women have to be minimised or eliminated in order to recruit and retain more women in construction which consequently will aid to make a change in the adversarial culture pertaining in the industry and will improve the equal opportunities of women. The benefit to the construction industry of reducing the perceived barriers to entry for females is a potentially radical shift towards collaborative relationships between project actors and between project actors and their stakeholders.

This chapter presents some findings concerning the lack of women in the construction industry with the aim to investigate how the construction industry can successfully recruit and retain professional women by looking at what lessons can be learned from other sectors. In this context this chapter intends to compare the construction industry with medicine and discovers how the medical sector has provided accessible careers for women. It also focuses on the culture of the medical sector, to establish the influence this has over the employment of female professionals.

Approach to the study

Women account for 9% of the construction workforce (CITB, 2003), of which 84% hold secretarial posts and 10% are employed in a professional capacity, in design and management areas. Therefore women carrying out construction-specific work make up less than 1% of the workforce; however, women at the professional level are amongst the most under-represented and, by implication, this is where scope for removing barriers is greatest; this chapter mainly focuses upon this area. The current status of professional women in construction has been investigated through literature survey and interviews. The methods used for the data collection and analysis are described below.

Literature review

Literature on women in construction has been critically examined in order to identify relevant issues and current practices, and to contextualise the research findings. The literature survey also considered medicine in order to identify its nature and the reasons behind the increased number of women in this sector. This enabled a comparison of the culture of the construction industry and medicine.

Interviews

Questions relating to factors that influenced the respondents' career decisions, career progression and barriers to progression, and their general experiences at work, were addressed. To address these questions, a series of interviews was conducted. Interviews were chosen as the most appropriate method for this study due to their appropriateness for capturing the experiences and meanings of the subjects in the everyday world, and as they allow subjects to convey to others their own situation from their own perspective and in their own words (Kvale, 1996). The type of interviews used were semi-structured; these have the advantage of being a 'halfway house' between the rigid layout of a structured interview and the flexibility and responsiveness of an unstructured interview (Moore, 2000). They allowed for the collection of both structured information and people's views and opinions, allowing spontaneity in the interviewer's questioning and the interviewee's response. The process of constructing and using qualitative research interviews can be divided into four steps: defining the research questions; creating the interview guide; recruiting participants; and carrying out the interviews (King, 2005). The research question was defined as 'what is the current status of professional women in construction and medical sectors, in terms of their recruitment and retention in the respective sectors?' Based on this research question, the interview guidelines were formulated separately for each industry sector. The qualitative research interview generally uses an interview guide, listing interview topics and suggesting probes which may be used to follow up responses and elicit greater detail from participants (King, 2005). The interview guide included piloting interviews within the team members, the guidelines being modified by adding probes and, in certain instances, additional topics. Research interviews were recorded.

Sample selection

The recruitment of participants to the interviews was carried out based on the intention of the research study, which was to compare the construction industry with medicine in terms of recruitment and retention of professional women. Accordingly, men and women working in a professional capacity within construction – including consultancy, contracting and client organisations – were identified and ten semi-structured interviews were conducted to gather information based on the interview guidelines. In medicine eight female medical officers, including senior house officers in psychiatry, senior house officers in anaesthesia, consultant paediatricians and general practitioners, were identified in order to compare their work with that of construction professionals, as they fall under the category of people working in a professional capacity.

Data analysis

The next stage was the analysis of the interviews, following the process outlined by Miles and Huberman (1994:10–12): *'qualitative analysis involves three activities: data reduction, data display, and conclusion drawing'*. The first stage of analysis identified is data reduction, which is the process of selecting, focusing and simplifying the interview transcripts by extracting the most relevant data for all of the questions and responses to additional probes. This process identified a number of issues under the major themes of the project. The second stage in the analysis process was the production of two data matrices – one for construction and one for medicine – by respondents and questions. This format allowed for comparison between the sectors. Patterns in the responses and relationships between the employment of women in construction and medicine were identified for detailed analysis and presentation.

Literature review

Current status of women in construction

The construction industry is one of the UK'S chief employers, employing over 2 million people and accounting for more than 1 in 14 of the total UK workforce (CITB, 2003). The role of the women in employment is changing radically in most societies, with women constituting just over half of the total workforce in Britain. Women constitute 9% of the construction industry, and this reflects their under-representation in the industry which fails to recruit and retain more women. Court and Moralee (1995) noted that the under-representation of women in construction only became an issue in the 1980s. In 1988, less than 7% of the full-time construction industry workforce in Britain was women. The Equal Opportunities Commission (1995) noted that women continued to be significantly under-represented in the primary sector (agriculture and energy and water), in most manufacturing, in transport and communications and, in particular, in the construction industry.

The issue regarding the under-representation of women in construction has been made more prominent recently, attracting government and industry-wide attention, due to the potential skill shortage facing the industry. The under-utilisation of human resources based on gender issues are of both social and economic concern. Therefore the UK government has been examining ways to encourage women into traditionally male-dominated jobs. Though researchers have focused on how to improve the participation of women in the construction workplace, the objectives seem to be aimed rather towards solving the labour resources crisis and skill shortages than improving equal opportunities for women (Agapiou, 2002). In contrast, certain benefactors, such as UK Resource Centre (UKRC) and Women in Science and Engineering (WISE), are mainly looking into the equal opportunities of women. These bodies provide immense support to attract more women into construction by means of providing training to women, educating women with the knowledge of construction career opportunities and providing mentoring. It is important to understand the impossibility of building a modern nation on the basis of both exclusion and inequality.

Despite the number of initiatives which have been introduced to solve the skill shortages and to improve the equal opportunities for women, the industry has failed to make significant progress in recruiting more women. It can be seen from Figure 10.1 that, over the years, the number of women working in the construction industry has remained constantly low.

Women in medicine

This section gives an overview of how the medical sector became accessible for female careers. The Royal College of Physicians (RCP) (2004) cites that the proportion of women medical students remained at 20–25% until 1968, and since then numbers increased steadily, exceeding 50% in 1991 and reaching 60% by 2004. A recent report by the British Medical Association (BMA, 2005) has now placed this figure at around 64% and still rising. This steady increase in female medical students is now noticeable in employment figures. Research by Finlayson *et al.*

Figure 10.1 Employee jobs in the UK construction industry, 1978–2006 (Office of National Statistics, 2007)

(2001) found that women now constitute a substantial part of the medical workforce in all grades, and the number in the medical specialities is also growing. However, it has been noted that the numbers of women working at consultant level and in positions of seniority remain comparatively low (BMA, 2004; RCP, 2004; Finlayson *et al.*, 2001). This could be due the time lag between women entering medical school and securing senior grades.

The perception of women in medicine started to change in the late 1960s due to the growing concern about the shortage of doctors and an over-reliance on doctors from overseas. As a result of these concerns there was an increase in the number of provincial schools, which meant women could be admitted without reducing the number of places available for men. Furthermore, in this period medical schools adopted the standard admissions procedures used by other universities, which took into account grades (A levels) rather than the 'old boy's network' approach (Elston, 1993).

The possible reasons for the increase of women in the medical profession could be the changes in the medical sector over the last century and changes in attitudes towards women in medicine. However a number of different reasons were presented by various medical authors, which are briefly discussed. A prominent factor behind the increase of women in medicine was the change in admissions procedures, as mentioned previously. This issue is significant, as women have traditionally done better than their male counterparts in A level examinations and nowadays girls are continually surpassing boys at all academic levels. Herbert (2004) identifies that unlike previous generations, modern women are in a position where they can be whatever they want to be, are working hard and achieving the high grades necessary for a career in medicine.

The increased entry of women to the medical sector was attributable to female high achievers being directed towards a career in medicine by 'a range of socialisers' such as school career's advisers and family members (Powell *et al.*, 2004). Similarly, studies by Jawitz *et al.* (2000) found that family members and particularly parents had a profound influence on career choice. Unlike construction, these socialisers were not perceived as having an image problem. Further, the career knowledge and the awareness of the medical sector in terms of its status and pay were comparatively high.

A further determining factor behind the increase of female doctors, as argued by Chidambaram (1993), is the need for doctors to be representative of their patients. During the 1980s the medical profession was under scrutiny following a spell of negative publicity focusing on cases of inappropriate behaviour and the functioning of male medical practitioners in relation to female patients. Thus it was perceived that women doctors would provide an 'equal and empathic' service to female patients.

However, studies have shown that women in medicine dominate the lower echelons of the profession and are under-represented in more senior and consultancy positions.

The effect of culture on the employment of women in construction and medicine

The expression 'culture', when discussed in terms of a corporation or organisation, is a complicated concept. This section discusses the effect of culture on the employment of women in construction and medical sectors.

Culture of the construction industry

In construction, culture causes further complications due to the size of the industry; not only is it necessary to consider the culture of the construction industry as a whole, but also the values of the several subsectors of the industry, such as the professions, house building, civil engineering, commercial contractors and property development.

Regardless of its different sectors and the disparity in values, the culture of the construction industry in general can be described as openly masculine (Gale, 1994). Similarly, Dainty *et al.* (2000) describe the construction industry as a male-dominated and threatening environment, with an engrained culture characterised by masculinity, conflict and crisis. Research by Bennett *et al.* (1999) corroborates this, discussing the male values of the industry. These include long working hours, working away from home, geographical instability and a highly competitive culture. In today's workplace these male values are generally considered extremely old-fashioned, but in construction, managers and professionals maintain these 'traditional' values and pass them on to others in the industry. Further Bagilhole *et al.* (2000) noted that the construction workplace has been described as among the most chauvinistic in the UK, with an extremely macho culture which is hostile and discriminatory against women. This results in gender-differentiated career opportunities, which have an inevitable consequence of high staff turnover of women in construction companies (Davidson and Cooper, 1992). Dainty *et al.* (2000) found that younger women became disillusioned with their career choice more rapidly than men, and sought to leave the industry early on in their careers. From this, it seems that the possibility of implementing a more open-minded culture is very slight.

This cultural problem together with the image of the construction industry is found to militate against the entry of women. Gale (1994) found that the image and culture of the construction industry strongly deter women from entering. Fielden *et al.* (2000) identified with this, citing the industry's poor image as a reason why so many people, regardless of gender, are uninterested in a career in construction. Almost a decade on from Gale's study (1994), research by Turrell *et al.* (2002) found that little had changed in the industry regarding its perceived image. The result of these views is that people, and women in particular, are less likely to enter the industry. This problem has been a concern for many years, once again brought to the forefront of industry attention by public figures such as Sir Michael Latham and Sir John Egan. More recently, the former Minister for Construction, Beverly Hughes, acknowledged the issue and stated that unless there was a step change in image and culture, the construction industry would face difficulties attracting new recruits to meet its long-term skills needs (Agapiou, 2002).

Culture of the medical profession

The culture of the medical profession in the past was not too dissimilar to that of the construction industry. Medicine was once, as construction is now, a male-dominated industry with similar factors to those seen in construction. These included long working hours, and evening, night and weekend work. Such practices are typically seen as unfavourable to women with family commitments.

As previously discussed, due to various reasons the medical profession has made considerable progress in the employment of women. Phillips (2004) cites that women make up 60% of all new doctors and identifies that in less than a decade they will outnumber men. This increase in women in medicine is significant considering how masculine both the profession, and the culture and values of the profession were. Like the construction industry, Dumelow and Griffiths (1995) stated that the 'male-orientated' organisational culture, the career structure, available opportunities and work practices of medicine were discriminatory towards the employment of women. Given the growing proportion of female doctors, it is clear that in general the medical profession has very few recruitment problems though there are still certain areas of medicine which fail to attract sufficient numbers of women.

Though the culture of medicine, in terms of male values, is seen as unfavourable to women, these issues clearly do not put women off entering a career in medicine. The reason behind this, in contrast to construction, is the positive image people have towards the medical profession. Furthermore, for many doctors, not just women, choosing to enter the profession is related to the high level of perceived prestige. Simon and Gick (1994) cited that this status originates from the public perception of doctors as bright, well educated and thoroughly trained individuals, which was particularly prevalent in the times when few were entering further education. On the other hand, Weger (1993) suggests that this prestige stems from the 'curing' aspect of the profession and the dependence that society has on doctors. The prestige of the medical sector has been maintained throughout the last few decades by the high grades required to enter medical school, the stature of the 'red brick' universities where these medical schools are situated and the financial rewards associated with the profession.

Summary

Many suggested that by recruiting more women into the construction industry, the culture could be changed to a certain extent. This corresponds with the Construction Industry Board's report on *Women and Men in Construction* (1996), which found that in many cases, women hold the skills essential for bringing about such cultural change. However, Gale (1994) noted that there was no evidence to prove that an increase in women in the industry would automatically change the culture, stating that those who have chosen a career in construction have chosen the culture of the profession and therefore have a stake in maintaining this culture.

However, a report by the RCP (2004) found that, unlike construction where part-time work is seen as impractical, the medical profession is more open to flexible working practices with a significant number of female practitioners, particularly those in more senior positions, working part-time. In medicine, support networks for women such as mentoring have grown in popularity as the number of women entering the profession has increased. They are now used to support women in medicine at all levels from mentoring during training, to 'women's groups' set up to offer encouragement to female practitioners.

Major findings of the primary research

Recruitment findings

This section is aimed at identifying possible initiatives to aid the recruitment of women to the construction industry and what lessons can be learnt from the medical profession. Analysis of this part of the interviews identified a number of key themes, discussed below, that attracted the women surveyed to their chosen career.

Family influences and childhood experiences were found to have an effect in stimulating an interest in male-dominated fields. Early childhood experiences and socialisation processes may well draw individuals to particular careers (Powell *et al.*, 2004). Women surveyed from construction felt they had very little exposure to the industry prior to starting work or construction education; the exceptions were two women with immediate family members working in the industry.

The opposite is the case in medicine. Just above 70% of the women surveyed in medicine had close family members working in the profession. Correspondingly, they all felt they were exposed to the profession.

> *I decided to pursue a career as a doctor when I was 16 after GCSEs because I had the family background in the medical professions; my mother and father both are doctors and I also had friends who were doing a medical degree as well.* (Female, 30, medicine)

> *I had a good awareness of the medical sector as several of my family members are doctors.* (Female, 31, medicine)

This raises a further point: the two women out of six in construction who had family with attachments to the industry were discouraged by their parents when they to choose construction as their career.

> *I decided to become a civil engineer when I was very young. I never wanted to do anything else. My father is a mechanical engineer; he told me not to be an engineer because he wasn't satisfied with the degree I did. He asked why didn't I become a lawyer or a doctor.* (Female, 27, construction)

This corresponds with the research of Dainty *et al.* (2000), who that found female entrants were less likely to have been advised to join the industry by friends or family with experience in construction, therefore those friends and family who had worked or were working in the industry were advising against or at least not encouraging women to pursue this career avenue. Where awareness and exposure to the industry were comparable to medicine, the effect was largely negative and reinforced the problems of culture and hence image in construction.

I have two daughters and I told them not to come into construction. (Male, 49, construction)

A further factor affecting career choice and, in turn, recruitment, which revealed mixed opinions, was the information and support the women had received from parents, career advisors and teachers. All of the women surveyed from the medical profession stated they had received good career advice from teachers and had very supportive parents and none had received any opposition to their career choice. Only one woman was not influenced by any of the career advisors, teachers or parents. She chose the medical profession based on her personal choice but her decision was supported by her family members.

I decided on a career as a doctor because I was really interested in caring for others and providing a service to them. In addition to that, I thought about the scope of progressing in the career in terms of academic and professional achievement, financial and social factors. (Female, 32, medicine)

I was given advice about a career in medicine and information about how to enter the profession. (Female, 26, medicine)

Whilst women surveyed from construction had faced little opposition, nearly 80% had received very little advice – 'silent discouragement' from parents and a lack of awareness and informed advice from careers advisors.

Before taking my A levels, I asked advice from the school tutor regarding my interest in construction. He just shouted at me as I didn't have the A levels at that time. (Female, 41, construction)

In school I was told not to do civil engineering. (Female, 26, construction)

When I saw the advert for quantity surveyors in the paper, I applied for the job and got an interview. Because I just finished my A levels, I went to the careers advice centre to find out what quantity surveyors did. She told me that I wouldn't get the job because it was a job for a man. Since I told her that I was going for the interview, she said to me that she would ring up and cancel it for me and also asked me to send some males along for the interview. (Female, 45, construction)

The final part of this section on recruitment concentrated on the opinions of these women on why the medical profession has become such a popular career destination for women, and why the initiatives developed by construction's professional bodies have had only limited success in attracting women to that industry. All of the women interviewed from the medical profession were unanimous as to why medicine has become popular for women: flexible training and work availability in general practice. One woman discussed the socially acceptable aspects of working in medicine compared to construction, and that medicine offered good opportunities for career progression and financial rewards.

However, the women in construction had mixed views regarding the initiatives aimed at attracting women to the industry. None of the women claimed to be influenced by such campaigns, being unsure about their effects and the image of the industry portrayed by them.

> *I hadn't seen a huge amount of these initiatives and my knowledge of such campaigns was limited to adverts with women in hard hats.* (Female, 36, construction)

> *I think the initiatives didn't clearly advise entrants on the discrimination and sexist attitudes women in construction face . . .* (Female, 30, construction)

This is in line with Dainty *et al.* (2000) who argued that women may not remain in the industry after education due to the inaccurate picture of the industry portrayed by recent recruitment initiatives. Their research found that women are more likely to be attracted to the industry by targeted recruitment campaigns, and they noted that women who had entered the industry due to such initiatives have a poor initial understanding of the culture of the industry and the inherent difficulties of working in a male-dominated environment.

Retention findings

This section focuses on the retention of women in their respective professions and the factors affecting retention. The construction industry seems to have problems retaining well educated female professionals. This part of the study aims at investigating the reasons why women leave the construction industry and why women doctors tend to stay in the medical profession.

Interview findings demonstrated mixed responses from the women working in construction with regard to whether they would leave, or have previously left the industry. Some would consider leaving the profession and some others had left the industry previously but had returned. In contrast, as predicted, none of the women surveyed from medicine had considered leaving the profession. The doctors with children had considered part-time working, but not leaving the profession. This raised the question as to the difference between these two sectors that caused women to leave one but not the other. The direction of the interviews then went on to consider this and an analysis of the results identified a number of key themes which are discussed below.

One of the main reasons that women leave the construction industry is difficulties associated with promotions. All of the women, except two leading women, interviewed from construction felt there was little possibility for promotion. However, since those had not been in the industry that long, it may be too early to know for sure whether this is truly the case; the time-lag factor was identified in the medical professions, yet all the women interviewed from the medical profession thought there was the possibility for progression. This is supported by the fact that the many women surveyed had already experienced promotion and had reached comparatively senior levels in their careers. Further, the two leading women who claimed there were no difficulties in the career progression and promotions were single and had no family commitments. They further acknowledged that it would not have been that easy if they had children.

Women surveyed from the construction industry did not feel they had faced any discrimination, nor any personal barriers that had hindered their progression. However, they claimed that this was not because they did not exist but because they had found ways to overcome any discrimination.

> *Well no, because I've always sort of gone round them. I think the discrimination tends to be more indirect, like the fact you'll get to site, and there'll be no ladies toilet. Or, you'll work with a 60 year old contracts manager or site agent that's been in the game for years and he thinks he knows it all, and thinks that women shouldn't be allowed and so he tries to make your life difficult rather than come out and say something to you.* (Female, 30, construction)

> *No, but I'm quite a determined person, I wouldn't take much notice anyway.* (Female, 32, construction)

> *There have been a couple of incidents where I've been told I'm just a woman. I think there's still that ethos about women and I think that's something that's going to be difficult to break.* (Female, 46, construction)

In contrast to the women in construction, 60% of the female doctors spoke openly about the discrimination and barriers they had faced at a personal level, in general, and in relation to career progression. One woman discussed that she had faced discrimination in the form of attitudes of patients.

> *Male doctors are considered more important . . . some people, not all people, would prefer to go to a male doctor.* (Female, 35, medicine)

Only one of the women stated that she had not faced any discrimination at all. However, she was the youngest woman surveyed and therefore may have entered after the profession had become more female orientated.

This theme raises the question of why women surveyed from construction were more unlikely to discuss discrimination openly than those from medicine. Could it be that they are educated into believing that such attitudes and values are the norm? Corresponding to this, Fielden *et al.* (2001) noted that despite legislation to remove gender discrimination, broader discriminatory behaviour culturally induces women in construction to expect to tolerate sexual discrimination whereas women in the medical profession clearly perceive discrimination and endeavour to confront or overcome it.

Data concerning mentoring from both industries revealed similarities. All of the women, irrespective of sector, felt that it was good to have a mentor, most having unofficial mentors rather than official ones. However women in construction argued that they experienced practical difficulties in finding a suitable person from whom to seek advice as they belonged to a minority group.

A further factor investigated through the interviews relating to the retention of women was what motivated these women in their careers. Surprisingly, all the construction professionals, both men and women, mentioned job satisfaction as their main motivation factor.

> *The fact that every day is different. Sometimes we have to react to problems and challenges. Satisfaction of doing a good job . . .* (Female, 41, construction)
>
> *I wanted to achieve really. I wanted to move forward in the industry. It gives me great self-satisfaction.* (Female, 27 – industry)
>
> *. . . Cheerful feeling when building roads and bridges.* (Male, 49 – industry)
>
> *Seeing people grow and develop. Seeing the built environment improve, it gives satisfaction . . .* (Male, 42 – industry)

All of the women surveyed from the medical profession mentioned financial rewards as their main motivator followed by status, job satisfaction and caring for people. However the financial rewards were not considered as important in construction as they were in medicine. This may reflect the fact that the construction industry is a low-paid industry compared to medicine.

Culture findings

Based on the respondents' opinions of the cultures of their respective industries, the cultures seem to be unfavourable to women working in both these professions.

However, the culture of the medical sector is now in a better position than it used to be due to ongoing changes in the profession such as the introduction of flexible training and implementing the Working Time Directive. Nearly 40% of doctors surveyed considered flexible working as the reason why medicine has become such a popular career choice for women. However, none of the respondents had fully utilised such arrangements; one tried to find part-time training but the waiting list for a part-time post was lengthy. This suggests that, although such practices are becoming more commonplace, they are still not that popular in hospital medicine. All of the doctors surveyed noted that such arrangements have still not been fully accepted and that they would be concerned about the effects such arrangements would have on their careers.

> *The shift system now present for doctors can help with this, but it is not 100% practicable. . . . you may have a 9am–5pm shift for example, but you cannot usually leave at 5pm; you have to finish all the necessary tasks for that day, as patients suffer if you do not.* (Female, 31, medicine)

All the women interviewed considered the culture of the construction industry to be adverse to women. When the question about flexible working was raised with the women in construction, they all responded that it would aid retention and make the industry more attractive to women. They further added that this would be an advantage for men in the industry as well since men also bear family commitments nowadays.

> *Even men will like that. The world is changing where men need to do half the work in terms of child caring and things like that. It is possible to adopt that kind of a culture in the industry but it will take a long time. It would have to come from the senior management level . . .* (Female, 35, construction)

> *. . . by introducing flexible working hours we can say that the construction industry will become more family friendly, rather than saying it is more female friendly.* (Female, 27, construction)

Though the culture of the medical professions is not too dissimilar to that of construction in terms of certain male values, it could be seen that the introduction of flexible working hours led the profession into a state where it is more receptive to women. The culture could include not only the male values and norms, but the way people are working, behaving and being in the sector. In this context the increased number of women in the medical sector is likely to facilitate positive cultural changes in those sectors in the future.

Conclusions

This chapter started with the proposition that increasing the numbers of females working in construction would help to change the culture from adversarial to collaborative. The chapter went on to deal with the outcomes of research work looking into recruitment and retention issues for women in the construction industry by looking at what lessons can be learned from another industry. Medicine was chosen for comparison.

The construction industry-based research found that the culture within construction seriously militates against women, both discouraging them from entering the industry and affecting their retention in the profession. This study also highlighted the similarities between the cultures and values of the medical and construction sectors. However, unlike the construction industry, medicine has made certain efforts to address its female recruitment problems by introducing flexible working hours. Further the increased number of women in the medical professions has created opportunities for more women to enter into the career by means of mentoring and networking.

From the findings it was clear that women who wanted to choose a career in medicine received positive advice and encouragement from their parents, teachers and career advisors. This was not the case with the women who wished to enter construction. There was a lack of knowledge and understanding about careers in construction and a poor image of industry prevailed.

Recommendations

Raising awareness of the construction industry and its various professional roles and career opportunities for women is important. Education has an important place in raising public awareness of the construction industry. Not only the students at school level, but the parents, teachers, industry employers and careers advisors also need to be educated. It is extremely important to give knowledge about the nature of construction industry's professional occupations and higher education routes to professional status and career opportunities in construction to school students considering a degree in construction. It was also suggested that young professionals should be encouraged to go into schools and talk about their careers in order to change incorrect perceptions that many people have towards the construction profession. Advice should be given to parents and teachers on how to encourage and support their daughters and students respectively who choose to have a career in construction. It is also imperative to prevent career advisors providing inaccurate and inadequate information on the construction industry, through organising training events for them where they can be educated with the career opportunities available for women in construction. As discussed earlier, the skill shortage is an ongoing problem in the UK construction industry at the moment. It is necessary to emphasise this problem to employers and let them know how the recruitment of more women might address this problem. Finally, raising awareness of the construction industry among the general public is very important as they are yet to realise the contribution of the built environment towards the quality of life.

Acknowledgement

This research was carried out with the financial support of the European Social Fund (ESF).

References

Agapiou, A. (2002) Perceptions of gender roles and attitudes toward work among male and female operatives in the Scottish construction industry. *Construction Management and Economics*, 20, 697–705.

Bagilhole, B.M., Dainty, A.R.J. and Neale, R.H. (2000) Women in the construction industry in the UK: A cultural discord? *Journal of Women and Minorities in Science and Engineering*, 6, 73–86.

Bennett, J.F., Davidson, M.J. and Gale, A.W. (1999) Women in construction: a comparative investigation into expectations and experiences of female and male construction undergraduates and employees. *Women in Management Review*, 14(7), 273–291.

British Medical Association (2004) *The Way Forward: Medical Women*. http://www.bma.org.uk, viewed April 2006.

Brooks, F. (1998) Women in General Practice: responding to the sexual division of labour? *Social Science and Medicine*, 47(2), 181–193.

Cartwright, S. and Cooper, C.L. (1994) *No Hassle! Taking the Stress out of Work*. Century Business, London.

Chidambaram, S.M. (1993) Sex stereotypes in women doctors' contribution to medicine: India. In: Riska, E. and Wegar, K. (eds.) *Gender, Work and Medicine: Women and the Medical Division of Labour*. Sage Publications, London.

Construction Industry Board (1996) *Tomorrow's Team: Women and Men in Construction*. Report of the CIB Working Group 8, Thomas Telford and Construction Industry Board, London.

Construction Industry Training Board (2003) *Construction Skills and Foresight Report*. http://www.citb.co.uk, viewed September 2004.

Court, G. and Moralee, J. (1995) *Balancing the Building Team: Gender Issues in the Building Professions*. Institute for Employment Studies/CIOB, University of Sussex.

Crompton, R. and Le Feuvre, N. (2003) Continuity and change in the gender segregation of the medical profession in Britain and France. *International Journal of Sociology and Social Policy*, 23(4/5), 36–58.

Dainty, A.R.J., Bagilhole, B.M. and Neale, R.H. (2000) A grounded theory of women's career under-achievement in large UK construction companies. *Construction Management and Economics*, 18, 239–250.

Dainty, A.R.J., Neale, R.H. and Bagilhole, B.M. (1999) Women's careers in large construction companies: expectations unfulfilled? *Career Development International*, 4(7), 353–357.

Davidson, M.J. and Cooper, C.L. (1992) *Shattering the Glass Ceiling. The Woman Manager*. Paul Chapman, London.

Dumelow, C. and Griffiths, S. (1995) We all need a good wife to support us. *Journal of Management in Medicine*, 9(1), 50–57.

Egan, Sir John (1998) *Rethinking Construction*. http://www.construction.detr.gov.uk./cis/rethink/index/htm, viewed July 2005.

Elston, M.A. (1993) Women doctors in a changing profession: the case of Britain. In: Riska, E. and Wegar, K. (eds.) *Gender, Work and Medicine: Women and the Medical Division of Labour*. Sage Publications, London.

Equal Opportunities Commission (1995) Job segregation linked to gender bias. *Equal Opportunities Review 60*, March/April, Equal Opportunities Commission, Manchester.

Fagan, C. and Burchell, B. (2002) *Gender, Jobs and Working Conditions in the European Union*. European Foundation for the Improvement of Living and Working Conditions, Dublin.

Fielden, S.L., Davidson M.J., Gale, A.W. and Davey, C.L. (2000) Women in construction: the untapped resource. *Construction Management and Economics*, 18, 113–121.

Fielden, S.L., Davidson, M.J., Gale, A.W. and Davey, C.L. (2001) Women, equality and construction. *Journal of Management Development*, 20(4), 293–304.

Finlayson, N., Lorimer, A.R. and Alberti, K.G.M.M. (2001) *Women in Hospital Medicine: Career Choices and Opportunities*. http://www.rcplondon.ac.uk, viewed June 2005.

Gale, A. and Cartwright, S. (1995) Women in project management: entry into a male domain? A discussion on gender and organisational culture – Part 1. *Leadership & Organisation Development Journal*, 16(2), 3–8.

Gale, A.W. (1994) Women in non-traditional occupations: the construction industry. *Women in Management Review*, 9(2), 3–14.

Greckol, S (1987) *Women into Construction*. National Association for Women in Construction, Toronto.

Green, E. (2005) *The Recruitment and Retention of Women in Construction: what lessons can construction industry learn from the medical profession with regards to the recruitment and retention of professional women?* Unpublished BSc quantity surveying dissertation, University of Salford, UK.

Gurjao, S. I. (2006) INCLUSIVITY:The changing role of women in the construction workforce. In: *Proceedings of the Construction in the XXI century: Local and*

Global Challenges – the Joint International Symposium of CIB Working Commissions, La Sapienza. Pietroforte, R., De Angelis, E. and Polverino, F. (eds) La Sapienza: School of Engineering, The University of Rome, 1–11.

Harris Research Centre (1989) *Report on Survey of Undergraduates and Sixth Formers*. Construction Industry Training Board, King's Lynn.

Heiligers, P.J.M. and Hingstman, L. (2000) Career preferences and the work life balance in medicine: gender differences among medical specialists. *Social Science and Medicine*, 50, 1235–1246.

Herbert, K. (2004) *Where have all the Men Gone?* http://www.studentbmj.com, viewed December 2005.

Jawitz, J., Case, J. and Tshabalala, M. (2000) Why not engineering? The process of career choice amongst South African female students. *International Journal of Engineering Education*, 16(6), 470–475.

King, N., (2005) Using interviews in qualitative research. In: Cassell, C. and Symon, G. (eds.) *Essential Guide to Qualitative Methods in Organisational Research*. Sage Publications, London.

Kvale, S. (1996) *Interviews: An introduction to Qualitative Research Interviewing*. Sage Publications, London.

Lambert, T.W., Goldacre, M.J., Parkhouse, J. and Edwards, C. (1996) *Career Destinations in 1994 of United Kingdom Medical Graduates of 1983: results of a questionnaire survey*. http://bmj.bmjjournals.com, viewed July 2005.

Lingard, H. and Francis, V. (2002) *Work-life Issues in the Australian Construction Industry: Findings of a Pilot Study*. Construction Industry Institute of Australia, Brisbane.

Lingard, H. and Lin, J. (2004) Career, family and work environment determinants of organizational commitment among women in the Australian construction industry. *Construction Management and Economics*, 22, 409–420.

Manser, M. and Thomson, M. (eds.) (1999) *Combined Dictionary Thesaurus*. Chambers Harrap Publishers Ltd., Edinburgh.

Miles, M.B. and Huberman, A.M. (1994) *Qualitative Data Analysis*, 2nd Edition. Sage Publications, London.

Moore, N. (2000) *How to do Research: The Complete Guide to Designing and Managing Research Projects*. Library Association Publishing, London.

Phillips, M. (2004) *Medicine Loses its Balance*. http://www.melaniephillips.com, accessed April 2005.

Powell, A., Baginhole, B.M., Dainty, A.R.J. and Neale, R.H. (2004) An investigation of women's career choice in construction. *Proceedings of the Association of Researchers in Construction Management Twentieth Annual Conference*, 1–3 September, Heriot Watt University.

Riska, E. and Wegar, K. (1993) Women physicians: a new force in medicine? In: Riska, E. and Wegar, K. (eds.) *Gender, Work and Medicine: Women and the Medical Division of Labour*, Sage Publications, London.

Royal College of Physicians, London (2004) *Briefing on Women in Medicine*. http://www.rcplondon.ac.uk/college/statements/briefing_womenmed.asp, accessed April 2006.

Simon, M. and Gick, P. (1994) *A Study of Physicians' Reactions to Health Care Reform*. http://www.drmsimon.com/articles/article5.htm, viewed June 2005.

Trinidad, C. and Normore, A.H. (2005) Leadership and gender: A dangerous liaison? *Leadership & Organization Development Journal*, 26(7), 574–590.

Turrell, P., Wilkinson, S.J., Astle, V. and Yeo, S. (2002) A gender for change: the future for women in surveying. *Proceedings of FIG XXII International Congress*, 19–26 April, Washington DC.

Weger, K. (1993) Conclusions. In: Riska, E. and Wegar, K. (eds.) *Gender, Work and Medicine: Women and the Medical Division of Labour*. Sage Publications, London.

White, B., O'Connor, D. and Garret, L. (1997) Stress in female doctors. *Women in Management Review*, 12(8), 325–334.

Whittock, M. (2002) Women's experiences of non traditional employment: is gender equality in this area a possibility? *Construction Management and Economics*, 20, 449–456.

Conclusion

Collaborative relationships in construction

Hedley Smyth and Stephen Pryke

It is intended that the conclusion to this book draws out the key points of the book, both for the future of industry and academic practice, and in so doing links it to the management of complex projects theme (Pryke and Smyth, 2006).

Aims and objectives

Project managers often comment that although intelligent thought and resources have been invested in ICT and similar investments made in structure and systems, still somehow the project coalition does not function as effectively and efficiently as it might. There are very serious financial implications of such shortfalls and failures, and the credibility and competitiveness of the industry as a whole is affected. A project manager responsible for a major demonstration project confided that, despite the promotional 'sizzle' created, the project had been 'one of *those* projects'. Neglect of the human and social dimension of managing projects and programmes is beginning to be addressed in more comprehensive ways. Value had for too long been considered as something added through inanimate tools and techniques, whereas it is people and in relationships that value is released.

In the **Introduction** we outlined the main *aim* as focusing upon the corporate–programme–project–client interfaces within a social network using a relationship approach. Flowing from this aim was a set of *objectives* that explored the limits of traditional practice and of recent 'best practice' of the reform and continuous improvement agendas. An exploration was provided which dealt with the link between corporate strategy for portfolio investment and programme management to support projects undertaken within frameworks and networks of relationships.

Past research and rationalisation of practice has artificially tried to shoehorn activities into constrained management silos. Collaboration, specifically through networks and frameworks, pushes beyond the organisational boundaries into networks of support that involve a broad range of actors in the identification of resources that are levered to help meet expectations and secured financial reward. Consequently, the corporate bodies on the client and construction side can harness relationships in frameworks, fitting these into the broader context of programme and project management.

Knowledge development

The book has contributed towards broadening the context and scope for developing a relationship approach, thus providing a point of departure from the confines of certain paradigms (see Pryke and Smyth, 2006; Smyth and Morris, 2007), of market structure and governance drivers. These market and governance issues are embedded in choices made about procurement strategies, including the important move towards relational contracting and the associated and inextricably linked opportunities and responsibilities associated with supply chain management.

There are learning points that arise from this overview. A more integrated approach is still necessary to embrace the totality and relationship of projects into their social environment; also, therefore, to place relationships in a central position, for example through transitioning from relational contracting to relationship management, to aid understanding of their contribution to adding organisational functionality from the corporate centre, from networks and across projects within frameworks.

Each chapter has made a valuable contribution showing different dimensions of this broader picture, the location of each contribution having been set out in Figures I.4, I.5 and I.6. Whilst each reader will draw their own lessons and knowledge from these chapters, we wish to draw out one particular theme that arises across a number of the chapters: the role of the public sector. The debate between private and public sector provision in mixed economies has been largely succeeded by service outsourcing, which in the public sector has led to privatisation and the use of PFI and PPP forms of procurement. The issue has become what form is the most efficient (and increasingly effective) form of delivery regardless of ownership. However, public sector organisations have more forcefully protected their organisational boundaries even under partnering and partnership arrangements. A number of the chapters have brought evidence to bear to directly and indirectly show that this is untenable for improving project outcomes in the short and long term – especially **Chapter 7** by Haigh, Amaratunga, Keraminiyage and Pathirage, **Chapter 8** by Volker and **Chapter 9** by London and Chen.

Just as it has been argued in this book and elsewhere (e.g. Pryke and Smyth, 2006; Smyth and Edkins, 2007) that the corporate centres of private organisations need to invest in relationships, systems and processes that deliver improvement (and deliver competitive advantage as core competencies and routines; see for example Hamel and Pralahad, 1994; Nelson and Winter, 1982; and in construction Smyth, 2011 forthcoming), the same holds true for public sector organisations to not only have consistent management of relationships but also to work across organisational boundaries.

Implications

The main implication has been to place projects and their management into a broader context. Much of the research into managing projects has primarily seen the 'project' – a narrowly defined categorisation that is primarily technical and managerial in conception. The absence or relegation of both the human and broader issues tends to overlook the socially constructed nature of the project and its context. This book did not set out to argue that a broader conception ought to be followed. The book is a challenge and hence can be used as a test of whether progress is made in the future on improving success. The take-up of the challenge may take a different form than description and analysis present here. That would of course be a measure of success too.

Traditional and reform modes of thinking and operations tend to point towards a failure to add value, improve performance and develop services (cf. Green, 2006). Forces of market competition or government regulation prevent improvement. Market forces have power by definition and power is exerted that at times will act as a constraint. Yet markets are socially constructed, and, like anything that is socially constructed, they can evolve and/or be transformed in ways that are not always comfortable, but in ways that go beyond current practices.

Future development

It has been our intention to stimulate discourse on projects and their management in ways that recognises the importance of relationships in a broader context that the project *per se*. Working in this broader context requires collaboration, typically across organisational boundaries, and leverage of resources from and through networks. We wish to see research that helps develop this, which includes refining and changing those things set down that may not stand the test of scrutiny. Therefore we recommend:

- The academic community needs to place social science approaches, especially relationships, at the heart of managing projects.
- The research community should use methodologies that recognise context and ensure that research is articulated in contextual terms in order for academics and practitioners to make connections between other areas of research and the subscribed collaborative areas of technical and management practice.
- The academic community needs to more deliberately acknowledge relationships as the source of social capital, hence for improved performance on programmes, projects and in project networks.

There are also some lessons for practice from the work presented here:

- Those responsible for procuring and managing complex projects and programmes of work need to collaborate with academic colleagues to understand the effectiveness of their networks and the roles and relationships embedded within their networks both retrospectively and when planning future programmes and strategies.
- The project community needs to give greater recognition to the need to manage relationships in frameworks and networks, requiring collaborative practices.
- The project community needs to give greater recognition to the role of programmes and portfolios, with attendant corporate investment and support in the management of projects.
- The industry bodies of knowledge need to take on board the broader spectrum.
- There needs to be greater recognition of the importance of managing relationships beyond the confines of the project: within organisations from the strategic centre to project operations, across multi-organisational teams and public–private interfaces.
- Contractors and sub-contractors currently lag behind clients in applying programme management to their portfolio of projects in ways that breakdown procurement silos.
- The importance of relationships in managing the market also needs addressing, particularly issues concerning market forces and the leverage of key actors, plus the development of social capital in market contexts.
- Managing relationships is a vast area; however, organisations may wish to focus upon particular forms of collaboration in order to build strength in particular areas of service provision.

References

Green, S.D. (2006) Discourse and fashion in supply chain management. In: Pryke, S.D. and Smyth, H.J. (eds.) *Management of Complex Projects: a Relationship Approach*. Blackwell, Oxford, 236–250.

Hamel, G. and Prahalad, C. (1994) *Competing for the Future*. Harvard Business School Press, Boston.

Nelson, R.R. and Winter, S.G. (1982) *An Evolutionary Theory of Economic Change*. Harvard University Press, Cambridge.

Pryke S.D. and Smyth, H.J. (2006) *The Management of Complex Projects: a Relationship Approach*. Blackwell, Oxford.

Smyth, H.J. (2011) *Core Competencies in Construction*. Blackwell, Oxford.

Smyth, H.J. and Edkins, A.J. (2007) Relationship management in the management of PFI/PPP projects in the UK. *International Journal of Project Management*, 25(3), 232–240.

Smyth, H.J. and Morris, P.W.G. (2007) An epistemological evaluation of research into projects and their management: methodological issues. *International Journal of Project Management*, 25(4), 423–436.

Index

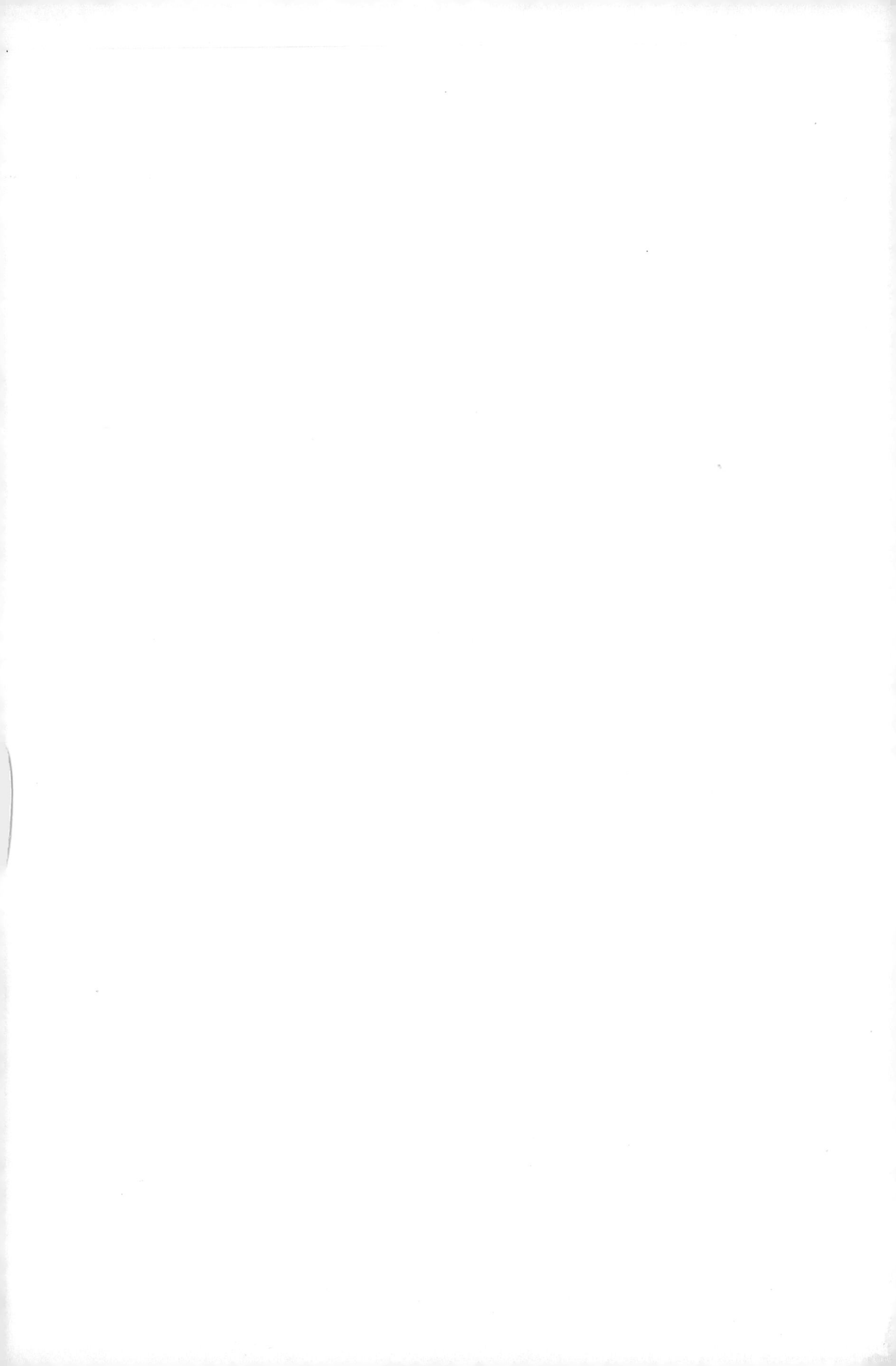